STABILITY OF LINEAR SYSTEMS: Some Aspects of Kinematic Similarity

This is Volume 153 in
MATHEMATICS IN SCIENCE AND ENGINEERING
A Series of Monographs and Textbooks
Edited by RICHARD BELLMAN, *University of Southern California*

The complete listing of books in this series is available from the Publisher upon request.

STABILITY OF LINEAR SYSTEMS:
Some Aspects of Kinematic Similarity

C. J. HARRIS
Department of Electrical and Electronic Engineering
The Royal Military College of Science
Shrivenham, Swindon, England.

and

J.F. MILES
Super Proton Synchrotron Division
European Organisation For Nuclear Research
1211 Genève 23, Switzerland.

1980

ACADEMIC PRESS

A Subsidiary of Harcourt Brace Jovanovich, Publishers
London New York Toronto Sydney San Francisco

ACADEMIC PRESS INC. (LONDON) LTD.
24/28 Oval Road,
London NW1

United States Edition published by
ACADEMIC PRESS INC.
111 Fifth Avenue
New York, New York 10003

British Library Cataloguing in Publication Data
Harris, C J
Stability of linear systems – (Mathematics in science and engineering).
1. System analysis
2. Stability
I. Title II. Miles, J F III. Series
003 QA402 78-75275
ISBN 0-12-328250-0

Printed in Great Britain

Preface

In spite of the considerable development in the last two decades of the state space approach to stability theory for linear time invariant systems the corresponding status of time varying and nonlinear systems is comparatively retarded. This apparent lack of maturity in the theory of variable coefficient and nonlinear differential equations can be ascribed to the need to derive the solutions of such systems before the structural properties of stability, controllability and observability can be ascertained. However for linear time invariant systems such properties can be determined directly (or indirectly through the algebraic approach of Laplace transforms) in terms of the coefficient matrices. It is the prime purpose of this book to identify classes of linear and nonlinear multivariable time varying coefficient differential systems whose stability can be characterised directly from their variable coefficient matrices by a suitable transformation, in much the same manner as linear time invariant systems. A secondary purpose of this book is to collect together and unify recent advances in linear stability theory and to highlight those results which are directly applicable to practical dynamic systems.

The book is self-contained and in Chapter One a complete review of mathematical preliminaries and definitions necessary throughout the book is given; the mathematically mature reader may omit this chapter without loss. This chapter covers various elements of functional analysis including linear transformations; matrix measures and their applications in estimating the bounds of solution to linear ordinary differential equations (Coppels inequality); inner product spaces and Fourier series, including Bessels inequality and Parsevals equation; and Cesaro sums and their associated Fejer kernels used in the approximation of real valued functions on bounded intervals.

As a prelude to the study of differential equations with almost periodic coefficients, the theory of almost periodic functions as a generalisation of pure periodicity is developed in Chapter Two. Properties such as Fourier series and Parsevals equation are established by analogy to the purely periodic case.

It is shown in an approximation theorem that to any almost periodic function there corresponds a sequence of trigonometrical polynomials which are uniformly convergent to the function. As many dynamical systems have spatially varying coefficients as well as time varying coefficients, the continuity, algebraic properties and Fourier series of almost periodic functions dependent upon a parameter are developed at length for later use in the context of asymptotic Floquet theory in Chapter Six.

Since the prime purpose of this book is the stability of linear dynamical systems, an introduction to ordinary linear differential equations and their properties is made in Chapter Three. Questions concerning the existence and uniqueness of solution are resolved via Picards method of successive approximations and the Gronwell-Bellman lemma which establishes bounds on solution. This latter result is important in stability studies since it yields an explicit inequality for the solution to an implicit integral inequality. Floquet theory describes linear ordinary differential equations with periodic coefficients; they occur in many theoretical and practical problems concerned with rotational or vibrational motion. It is shown that there exists a nonsingular periodic transformation of variables which transform linear periodic coefficient differential systems into constant coefficient systems; this form of Liapunov Reducibility or Kinematic Similarity is clearly important in stability studies.

The question of structural invariants, such as stability, under Kinematic Similarity are discussed together with the necessary and sufficient conditions for Kinematic Similarity for a variety of coefficient matrices in Chapter Four. Special emphasis is given to systems whose coefficient matrices commute with their integral; for such systems it is shown that the state transition matrix and Liapunov transform are readily computed and that unstable time invariant systems can be stabilised by time varying control laws.

Chapter Five is devoted entirely to the establishment of necessary and sufficient conditions for the stability of nonstationary differential equations with particular reference to linear systems with periodic and almost periodic coefficients. The theory of exponential dichotomy illustrates the danger of determining system stability based only on the characteristic values of time dependent coefficients. A more restrictive, but less conservative theory based upon the asymptotic behaviour of characteristic values for the stability of linear nonstationary systems is developed via matrix projection theory.

The investigation of Kinematic Similarity is taken up again in Chapter Six in the context of linear differential equations with almost periodic coefficient matrices and those dependent upon a parameter. Analogues with Floquet theory are identified and conditions for Kinematic Similarity are established via the characteristic exponents of the almost periodic coefficient matrices and the characteristic values of the transformed system.

By way of example, Chapter Seven contains a collection of

practical applications of linear differential systems with variable coefficients; these demonstrations include a pendulum with moving support, parametric amplifiers, columns under periodic axial load, electrons in a periodic potential, spacecraft attitude control and a detailed study on the beam stabilisation of a proton beam in an alternating gradient proton synchrotron.

This book is the result of a collaborative effort between the authors at the University of Manchester Institute of Science and Technology, Oxford University, European Organisation for Nuclear Research (CERN) and the Royal Military College of Science, and the authors wish to acknowledge their debt to these institutions for their support and the provision of facilities to carry out this work. Finally, personal thanks are given to Miss Lucy Brooks whose excellent typing turned an untidy manuscript into the final version of this book.

July 1980

C. J. Harris
J. F. Miles

CONTENTS

Chapter 1

MATHEMATICAL PRELIMINARIES

1.1 Metric Spaces

Metric spaces are fundamental in functional analysis since they perform a function similar to the real line R in ordinary calculus. A metric space is a set X with a metric defined on it. The metric associates any pair of elements x,y of X with a distance function d(x,y) which is essentially a generalisation of the distance between two points in a Euclidean plane. The metric space is defined axiomatically by:

Definition 1.1: *Metric space*

A *metric space* is a pair (X,d) where X is a set and d a *metric* on X, that is a function defined on the Cartesian product $X \times X$ such that for all $x,y \in X$ we have:

m1. d(x,y) is real valued, finite and non-negative

m2. $d(x,y) = 0$ if and only if $x = y$

m3. $d(x,y) = d(y,x)$

m4. $d(x,y) \leq d(x,z) + d(z,y)$, the triangle inequality

A *subspace* $(Y,\tilde{d})$ of (X,d) is obtained if we take a subset $Y \subset X$ and restrict d to $Y \times Y$, that is $\tilde{d} = d/_{Y \times Y}$ (which is known as the *induced* metric on Y by d).

Example 1

a) On the real line R the metric is $d(x,y) = |x-y|$

b) On the Euclidean plane $E^2 = R^2$, the Euclidean metric is

$d(x,y) = ((\alpha_1-\beta_1)^2 + (\alpha_2-\beta_2)^2)^{\frac{1}{2}}$ where $x = (\alpha_1,\alpha_2)$, $y = (\beta_1,\beta_2)$. Alternatively $d_1(x,y) = |\alpha_1-\beta_1| + |\alpha_2-\beta_2|$; this second metric illustrates that a given set X can have various metric spaces simply by choosing different metrices.

The generalisation of E^2 to the complex n-Euclidean space or unitary space C^n, is the space with the set of all ordered n-tuples of complex numbers $x = (\alpha_1,\ldots,\alpha_n)$, $y = (\beta_1,\ldots,\beta_n)$ with metric $d(x,y) = [|\alpha_1-\beta_1|^2 + \ldots + |\alpha_n-\beta_n|^2]^{\frac{1}{2}}$.

c) Consider the set X of all real valued continuous functions $x(t)$ on t over the closed interval $I = [a,b]$ with metric $d(x,y) = \max_{t\in I} |x(t)-y(t)|$, the space (X,d) in this case is called the *function space* C[a,b].

d) *Sequence spaces* ℓ^p. As a set X take all bounded sequences of complex numbers $x = (\alpha_1,\alpha_2,\ldots) \equiv x(\alpha_j)$ such that $|\alpha_1|^p + |\alpha_2|^p + \ldots = \sum_{j=1}^{\infty} |\alpha_j|^p < \infty$ for $\infty > p \geq 1$. If we now define the metric d(x,y) by $d(x,y) = [\sum_{j=1}^{\infty} |\alpha_j-\beta_j|^p]^{\frac{1}{p}}$ for fixed p and $y = y(\beta_j)$, $\sum_{j=1}^{\infty} |\beta_j|^p < \infty$, then in the metric space $\ell^p \equiv (X,d)$ each element $x = x(\alpha_j)$ converges for fixed p $(\infty > p \geq 1)$. For the special case of $p = \infty$, the metric on this set X is given by $d(x,y) = \sup_{j\in N} |\alpha_j-\beta_j|$, where $N = \{1,2,\ldots\}$ and $y = y(\beta_j)$.

A particularly important example of the ℓ^p space is when $p = 2$ in which case we have the *Hilbert* space with metric $d(x,y) = (\sum_{j=1}^{\infty} |\alpha_j-\beta_j|^2)^{\frac{1}{2}}$ and the following so-called Cauchy-Schwartz inequality holds

$$\sum_{j=1}^{\infty} |\alpha_j\beta_j| \leq (\sum_{k=1}^{\infty} |\alpha_k|^2)^{\frac{1}{2}} (\sum_{m=1}^{\infty} |\beta_m|^2)^{\frac{1}{2}} \tag{1.1}$$

A generalisation of this is possible for $p \geq 1$ if a q is such that $\frac{1}{p} + \frac{1}{q} = 1$, then we have Holder's inequality

$$\sum_{j=1}^{\infty} |\alpha_j \beta_j| \leq (\sum_{k=1}^{\infty} |\alpha_k|^p)^{\frac{1}{p}} (\sum_{m=1}^{\infty} |\beta_m|^q)^{\frac{1}{q}} . \qquad (1.2)$$

Rather than use products of elements of X, if we use sums in the sequence spaces for $p \geq 1$ we have Minkowski's inequality

$$(\sum_{j=1}^{\infty} |\alpha_j + \beta_j|^p)^{\frac{1}{p}} \leq (\sum_{k=1}^{\infty} |\alpha_k|^p)^{\frac{1}{p}} + (\sum_{k=1}^{\infty} |\beta_k|^p)^{\frac{1}{p}} \qquad (1.3)$$

for $x = x(\alpha_j) \varepsilon \ell^p$, $y = y(\beta_j) \varepsilon \ell^p$ and $p \geq 1$.

Metric spaces are a special class of *topological* spaces which are characterised by open sets in a space X. Since the important analysis concepts of continuity of transformations and convergence of sequences can be defined for general spaces in terms of open sets completely independently of a metric.

Consider a given metric space (X,d) we now discuss some of its topological properties:

Definition 1.2

Given a $x_o \varepsilon X$ and a real number $r > 0$ then we have the following sets:

$B(x_o,r) = \{x \varepsilon X: d(x,x_o) > r\}$ is an *open ball*,

$\tilde{B}(x_o,r) = \{x \varepsilon X: d(x,x_o) \leq r\}$ is a *closed ball*,

$S(x_o,r) = \{x \varepsilon X: d(x,x_o) = r\}$ is a sphere;

x_o is called the *centre* and r the *radius*. Clearly

$S(x_o,r) = \tilde{B} - B.$

An open ball of radius ε is called an *ε-neighbourhood* of x_o $(\varepsilon > 0)$.

Definition 1.3

A subset Y of X is said to be *open* if it contains an ε-neighbourhood about each of its elements. A subset Y of X is said to be *closed* if its complement Y^c in X is open, that is $Y^c = X - Y$ is open.

It is not difficult to show that the collection of all open subsets of X, called J has the following properties:

t1. the null or empty set $\theta \varepsilon J$, $X \varepsilon J$,

t2. $\bigcup_i X_i \in J$, for $X_i \in J$,

t3. $\bigcap_i X_i \in J$, for $X_i \in J$ and i finite.

The space (X,J) is called a *topological space* with the set J a *topology* for X; clearly a metric space is a topological space.

Open sets also play an important role in the concept of continuous mappings on metric spaces.

Definition 1.4: *Continuous mappings*

Let $X = (X,d)$ and $Y = (Y,\tilde{d})$ be metric spaces. A mapping $f:X \rightarrow Y$ is said to be *continuous* at $x_o \in X$ if for every $\varepsilon > 0$ there is a $\delta(\varepsilon) > 0$ such that $\tilde{d}(fx,fx_o) < \varepsilon$ for all x such that $d(x,x_o) < \delta(\varepsilon)$. f is said to be continuous if it is continuous at every point of X.

The mapping $f:X \rightarrow Y$ is *uniformly continuous* if in the above definition $\delta = \delta(\varepsilon)$ is independent of ε.

Example 2

If the mapping $f:X \rightarrow Y$ is represented by the matrix $A = \{a_{ij}\}$ such that $y = Ax$ with $X = R^n$, $Y = R^m$ real Euclidean spaces, then

$$d(y,y_o)^2 = \sum_{i=1}^{m} \left| \sum_{j=1}^{n} a_{ij}(x_j - x_{oj})\right|^2$$

$$\leq \sum_{i=1}^{m} \left(\sum_{j=1}^{n} |a_{ij}|^2\right)\left(\sum_{j=1}^{n} |x_j - x_{oj}|^2\right)$$

$$\leq \left(\sum_{i,j=1} |a_{ij}|^2\right) d(x,x_o) \equiv \gamma^2 d(x,x_o)^2 \qquad (1.4)$$

Thus selecting $\gamma\delta = \varepsilon$ for any positive ε, then by inequality (1.4) and definition **1.4** the matrix mapping f is uniformly continuous.

Continuity of a mapping in terms of open sets is contained in the following theorem whose proof utilises the above definition of open sets and continuity:

Theorem 1.1

A mapping f of metric space X into a metric space Y is

continuous if and only if the inverse image of any open set of Y is an open set of X.

In a similar fashion open sets can be used to define convergence for a sequence in a topology (X,J).

Definition 1.5: *Convergence of sequences*

A sequence $\{x_n\}$ in the topological space (X,J) converges to $x \in X$ if and only if for large n, x_n is in every open set that contains x.

We now consider two more related topological concepts. Let $Y \subset X$, a metric space, then $x_0 \in X$ (which may or may not be an element of Y) is called an *accumulation* or limit point of Y if every neighbourhood of x_0 contains at least one point $y \in Y$ distinct from x_0. The set $\overline{Y}$ consisting of the points of Y and the accumulation points of Y is called the *closure* of Y. That is for the topological space (X,J), $Y \subset X$, the closure of Y is

$$\overline{Y} = \cap\{C \supset Y\text{: } C \text{ closed in } (X,J)\}.$$

$\overline{Y}$ is a closed subset of (X,J) containing Y (in fact the smallest); in addition if $\overline{Y} = Y$ then Y is closed in (X,J). The concepts of set closure and closed sets enables us to make the following equivalent statements about the mapping $f:X \to Y$ for (X,J) and (Y,U) topological spaces;

(i) $f:(X,J) \to (Y,U)$ is continuous.

(ii) $f^{-1}(C)$ is closed in (X,J) for all closed C in (Y,U),

(iii) $f(\overline{N}) \subset \overline{f(N)}$ for all $N \subset X$.

Definition 1.6

A subspace N of a metric space X is said to be dense in X if $\overline{N} = X$.

This means that any $x \in X$ can be approximated by some element y of N with as small an error as we wish so that $d(x,y) \leq \varepsilon$ for arbitrary ε. All linear normed space have dense subsets, but for the approximation it is useful if a countable dense subset can be found.

Definition 1.7

A metric space X is *separable* if it has a countable subset which is dense in X.

Obvious examples of separable metric spaces are the real line R, complex plane and the space $\ell^p (\infty > p \geq 1)$, however the ℓ^∞ space is not separable since it contains uncountably many sequences each contained within non-intersecting balls.

Since metric spaces are special classes of topological spaces the definition of convergence in a metric space can be simplified to:

Definition 1.8

A sequence $\{X_n\}$ in the metric space $X = (X,d)$ is said to *converge* if there is a sequence $x \in X$ such that $\lim_{n\to\infty} d(x_n,x) = 0$ and $x_n \to x$.

Therefore if $X = (X,d)$ is a metric space a convergent sequence in X is bounded and its limit is unique; also if $x_n \to x$ and $y_n \to y$ in X as $n \to \infty$ then $d(x_n,y_n) \to d(x,y)$ as $n \to \infty$. The convergence of sequences in a metric space is closely connected with the continuity of a mapping between two metric spaces (X,d) and $(Y,\tilde{d})$, since the mapping $f:X \to Y$ is continuous at a point $x_o \in X$ if and only if $x_n \to x_o$ implies that $fx_n \to fx_o$. We note that in ordinary calculus a sequence $\{x_n\}$ converges if and only if it satisfies a Cauchy convergence criterion, similarly for metric spaces we have:

Definition 1.9

A sequence $\{x_n\} \in X = (X,d)$ is said to be a *Cauchy sequence* if for every $\xi > 0$ there is a $N(\xi)$ such that $d(x_m,x_n) < \xi$ for every $m,n > N$. Also if every Cauchy sequence in X converges then the metric space is *complete*.

Whilst every convergent sequence in a metric space is a Cauchy sequence, not all metric spaces are complete. This is unfortunate since a large number of results in the theory of linear operators depend upon the completeness of the corresponding spaces.

Example 3

The real line R and complex plane are examples of complete metric spaces, other important metric spaces that when complete are R^n, C^n, ℓ^∞ and ℓ^p. A particularly important complete metric space for our purposes is the function space $C[a,b]$ for $[a,b] \varepsilon R$; in addition the convergence $x_n \to x$ in this metric space is *uniform*, and so the metric $d(x,y) = \max_{t\varepsilon[a,b]} |x(t)-y(t)|$ is called the *uniform metric*.

Examples of incomplete metric spaces are the rational line Q composed of all rational numbers and the set of all continuous valued functions with metric

$$d(x,y) = \left(\int_a^b |x(t)-y(t)|^2 dt\right)^{\frac{1}{2}} \qquad \text{defined on} \quad [a,b] \varepsilon R.$$

We note that in this example the space of continuous valued functions defined on the interval $[a,b]$ has had two metrics defined on it, however only one of the metric spaces is complete.

1.2 Normed Metric Spaces

The most important metric spaces are vector spaces with metrics defined by a norm which generalises the concept of the length of a vector in a three-dimensional space. A mapping from a normed space X into a normed space Y is called an *operator*; also if Y is a scalar field then this mapping is called a *functional*. Of particular importance in the sequel are bounded linear operators and functionals since they are both continuous. Indeed a linear operator is continuous if and only if it is bounded.

Consider the field K of scalar real or complex numbers:

Definition 1.10: *Vector space*

A vector space X (or linear space) over a field K is a non-empty set of elements $x,y,\ldots$(vectors) which satisfy the algebraic operations

v1. $x + y = y + x$

v2. $x + (y+z) = (x+y) + z$

v3. $x + 0 = x, \quad x + (-x) = 0$

v4. $\alpha(\beta x) = (\alpha\beta)x,$ where α,β are scalars

v5. $\alpha(x+y) = \alpha x + \beta y, \quad (\alpha+\beta)x = \alpha x + \beta x.$

Example 4

Examples of linear vector spaces are R^n, C^n the n-Euclidean real and complex spaces, the function space $C[a,b]$, and ℓ^2.

A linear *subspace* of a vector space X is a non-empty subset $Y \subset X$ such that for all $y_1, y_2 \in Y$ and all scalars α,β, $\alpha y_1 + \beta y_2 \in Y$. Linear subspaces have the property that they all contain the zero element. A special subspace of X is the *improper subspace* $Y = X$.

A linear combination of vectors $x_1, x_2, \ldots, x_m$ of a vector space X is $\alpha_1 x_1 + \alpha_2 x_2 + \ldots + \alpha_m x_m$ for all α_i scalars. For any non-empty subset $N \subset X$, the set of all combinations of vectors N is called the *span* of N, which is also a subspace of X. The set of vectors $x_1, \ldots, x_r, \in N \subset X$ for $r \geq 1$ are said to be *linearly independent* if

$$\alpha_1 x_1 + \ldots + \alpha_r x_r = 0$$

only if $\alpha_1 = \alpha_2 = \ldots = \alpha_r = 0$.

A vector space X is said to be of *dimension n* (and finite), i.e. $\dim X = n$, if X contains a linearly independent set of n-vectors whereas any other set of (n+1)-vectors in X are *linearly dependent*. If $\dim X = \infty$ we say that the vector space X is infinite dimensional.

Clearly the vector spaces $C[a,b]$ and ℓ^2 have $\dim X = \infty$, whereas R^n and C^n are finite dimensional with $\dim X = n$. If $\dim X = n < \infty$, then a set of n-linearly independent vectors in X is called a *basis* for X and every vector $x \in X$ has a unique representation as a linear combination of the basis vectors. Clearly every linear vector space has a basis, and that all finite dimensional spaces are separable.

To combine the algebraic concepts of linear vector spaces and the geometric concepts of a metric we need normed linear vector spaces or simply normed spaces:-

Definition 1.11: *Normed spaces*

A norm on a (real or complex) vector space X is a real valued function on X whose value is denoted by $||x||$, with the properties:

n1. $||x|| \geq 0$,

n2. $||x|| = 0 \iff x = 0$,

n3. $||\alpha x|| = |\alpha| \, ||x||$,

n4. $||x+y|| \leq ||x|| + ||y||$, $\implies$ $| \, ||y|| - ||x|| \, | \leq ||y-x||$,

for $x, y \in X$ and α any scalar.

A norm defines on X a metric $d(x,y) = ||x-y||$ $(x,y \in X)$ and is called the metric *induced* by the norm. If the condition **n2** does not hold then we call $||x||$ a *semi-norm*. Condition **n4** implies that $x \mapsto ||x||$ is a continuous mapping of the normed vector space $X = (X, ||\cdot||)$ into the real line R. A *Banach space* is a complete normed vector space. We have already shown that the Euclidean spaces R^n and C^n, spaces ℓ^p $(p=1,\ldots\infty)$ and the function space C[a,b] are complete, in addition their respective metrics all satisfy the conditions of a norm and therefore they are all Banach spaces.

Example 5: The L^p spaces

We say that the function $f:R \to R$ is integrable if and only if f is integrable over the bounded interval [a,b]. Consider the space L^p of all (Riemann) integrable functions $f:R \to R$ such that f^p is integrable on some interval $[a,b] \in R$ for any f in L^p. We define the norm on this space by

$$||f||_p = \left(\int_a^b f(t)^p \, dt \right)^{\frac{1}{p}} \tag{1.5}$$

So if $\alpha \in R$ and $f,g \in L^p$ then $\alpha f \in L^p$ and from the inequalities

$$|f+g| \leq 2 \max \{|f|,|g|\}$$

$$|f+g|^p \leq 2^p \max\{|f|^p,|g|^p\}$$

it follows that $(f+g) \in L^p$ and therefore L^p is a linear space. The norm (1.5) on L^p satisfies all the conditions of a norm except that $||f|| = 0$ if and only if $f = 0$ *almost* everywhere. This is because a Cauchy sequence in L^p does not have its limit in this space, so that L^p is not complete and cannot therefore be a Banach space. However it is well known that every metric space has a completion which is unique up to an isometry. The completion of the L^p space is achieved by using Lebesgue integration - a generalisation of Riemann integration. (For simplicity, readers unfamiliar with Lebesgue theory should consider that all functions $f:R \to R$ are piecewise continuous.) We define a new space $L^p[a,b]$ whose elements are equivalence classes $\hat{f}$ of functions in L^p according to the equivalence relation,

$$f \sim g \quad \text{if and only if} \quad f = g \quad \text{almost everywhere.}$$

And define

$$||\hat{f}|| = ||g||_p = \left(\int_a^b |g(t)|^p \, dt\right)^{\frac{1}{p}} \tag{1.6}$$

with $\int_0^\infty |g(t)|^p \, dt < \infty$. Where g is any representative of the equivalence class $\hat{f}$. The linear space structure on $L^p[a,b]$ is defined in terms of representations from L^p such that

$$(\hat{f}+\hat{g}) = (f+g)^\wedge$$

$$\alpha\hat{f} = (\alpha f)^\wedge$$

In this manner the norm on $L^p[a,b]$ has the desired property that $||\hat{f}|| = 0$ if and only if $\hat{f} = \hat{0}$.

It is not difficult to show that if $f,g \in L^p$ $(p \geq 1)$ then Minkowski's inequality holds and in particular if $f \in L^p[a,b]$ $g \in L^q[a,b]$ then $fg \in L^1[a,b]$ and Holder's inequality

$$\int_a^b |fg|dt \le \|f\|_p \|g\|_q \quad \text{holds.}$$

In $L^p[a,b]$ spaces there are at least three kinds of convergence of a sequence $\{f_n\}$, with $f_n \in L^p[a,b]$ (we denote $L^p[a,b]$ by L^p in the sequel).

(i) If $\{f_n(t)\}$ converges to $f(t)$ $(f:R \to R)$ then $\{f_n\}$ converges *pointwise* to f.

(ii) If $\{f_n(t)\}$ converges to $f(t)$ for almost all t, then $\{f_n\}$ converges pointwise *almost everywhere* to f. The limit function in both these cases may or may not belong to L^p.

(iii) If $\{f_n(t)\}$ converges to f in L^p, if $f \in L^p$ and $\|f_n - f\|_p \to 0$ as $n \to \infty$, this is called *strong convergence* or *convergence in the mean* (of order p).

If $\{f_n\}$ converges strongly to both f and g then $f = g$ almost everywhere. However in the case of L^p spaces, pointwise convergence does not imply strong convergence and strong convergence does not imply pointwise convergence. For example take the sequence

$$f_n(t) = \begin{cases} n & \text{for } t \text{ in } (0,n^{-1}) \\ 0 & \text{otherwise} \end{cases}$$

the sequence converges pointwise to zero, however

$$\int |f_n|^p dt = n^{p-1} \quad \text{so that} \quad \|f_n\|_p = n^{(p-1/p)} \to \infty$$

as $n \to \infty$ for $p > 1$. However in the Euclidean R^n and C^n spaces strong convergence and pointwise convergence are equivalent.

Finally we note that for the $L^p[a,b]$ space, the monotone convergence theorem shows that this metric space is complete and is therefore a Banach space.

Finite dimensional normed spaces are much simpler than infinite dimensional spaces; for example every finite dimensional subspace Y of a normed space X is complete and closed in X.

Another important consequence of finite dimensional vector spaces X is that all norms on X lead to the same topology for X irrespective of the choice of norm on X. This leads to:

Definition 1.12: *Equivalent norms*

A norm $||x||$ on a vector space X is said to be *equivalent* to a norm $||x||_o$ on X if there are positive numbers α, β such that for all $x \varepsilon X$

$$\alpha ||x||_o \leq ||x|| \leq \beta ||x||_o.$$

We note that the same α, β must work for all $x \varepsilon X$. On a finite dimensional vector space X any norm $||x||_r$ is equivalent to any other norm $||x||_s$; this result implies that convergence of a sequence in X does not depend upon the choice of norm on that space. The same conclusion also holds for continuity and boundedness, that is equivalent norms define identical topologies. However in applications some norms are preferable since they give sharper results in say the estimation of eigenvalues (characteristic values) of matrix operators.

An important topological concept is that of compactness:-

Definition 1.13: *Compact spaces*

A metric space X is said to be *compact* if every bounded sequence in X has a convergent subsequence.

A general property of compact sets is that a compact subset N of a metric space X is closed and bounded (that is $\sup_{x \varepsilon X} ||x|| < \infty$). In addition for a finite dimensional normed space X any subset $N \subset X$ is compact if and only if N is closed and bounded (this is not true for infinite dimensional spaces). An interesting conclusion can be made about a normed space X if it has a compact subset $N = \{x: ||x|| < 1\}$, in which case X is finite dimensional. In connection with continuous mappings $f:X \to Y$ from a metric space X to a metric space Y, the image of a compact subset N of X under the transformation f is compact and this mapping achieves a maximum and a minimum at some points of N if $Y = R$.

For regular continuous valued functions defined on compact metric spaces an important theorem used in establishing the existence and uniqueness of solution to differential equations is the Arzela-Ascoli theorem, but first we need to define equicontinuiuty:

Definition 1.14: *Equicontinuous sequences*

A sequence $\{x_n\}$ in C[a,b] is *equicontinuous* if for every $\xi > 0$ there is a $\delta(\xi) > |t_1 - t_0| > 0$ such that $|x_n(t_1) - x_n(t_0)| < \xi$ for all x_n and all $t_1, t_0 \in [a,b]$.

From this definition each x_n is uniformly continuous on [a,b] and δ does not depend upon n.

Theorem 1.2: *(Arzela-Ascoli)*

A bounded equicontinuous sequence $\{x_n\}$ in the compact metric space C[a,b] has a subsequence which converges uniformly in the norm of C[a,b].

Note that the sequence $\{x_n(t)\}$ is uniformly bounded, that is

$$\sup_n \sup_{t \in [a,b]} |x_n(t)| < \infty.$$

1.3 Contraction Mappings

The contraction mapping (or Banach fixed point) theorem is a very important result which is used in establishing the existence and uniqueness of solution to nonlinear differential equations; equally it has played an important role in developing practical stability criterion, such as the circle criterion, for multivariable nonlinear systems.

Definition 1.15: *Contraction mapping*

For a metric space (X,d) the mapping $f: X \to X$ is a *contraction mapping* if there is a real number k $(0 \leq k < 1)$ such that $d(fx,fy) \leq kd(x,y)$ for all $x,y \in X$.

From definition 1.4 the above implies that the mapping f is uniformly continuous. The operator f is called a contraction since the images of any two elements x and y are nearer to each other than x and y are.

Theorem 1.3: *Contraction Mapping*

Let (X,d) be a complete metric space and let $f:X \to X$ be a mapping such that

$$d(fx,fy) \leq k\,d(x,y)$$

holds for a fixed constant k $(0 \leq k < 1)$ and for all $x,y \in X$. Then there exists exactly one $x_o \in X$ such that $fx_o = x_o$ and that for any $x \in X$ the sequence $\{x_n\}$ $(n = 1,2,\ldots,\infty)$ defined by $x_{n+1} = f\,x_n$ converges to x_o as $n \to \infty$; moreover $d(x_o,x_n) \leq \frac{k^n}{(1-k)}\,d(fx_o,x_o)$.

The unique point x_o is called a *fixed point* of the operator f, since it is fixed for every f applied to the space X. The last inequality of theorem 1.3 provides an estimate of the rate of convergence of the given sequence to the fixed point, and is particularly useful in evaluating numerical algorithms that compute the solution to integral and differential equations. The theorem holds globally since the contraction condition holds for all x,y in X; also it holds for all Banach spaces if (X,d) is a normed linear vector space. The proof of the theorem follows from showing that the sequence $\{x_n\}$ $(n = 1,2,\ldots,\infty)$ is a Cauchy sequence and since X is complete, converges to a limit x_o in X. The definition of uniform continuity shows that the mapping f is uniformly continuous, therefore this limit x_o is invariant under f, from which the contraction condition of the theorem shows that x_o is unique in X. A corollary to theorem 1.3 is:-

Corollary 1.3

If (X,d) is a complete metric space and if the mapping $f:X \to X$ is such that f^r has a contraction for some $r > 0$, then the mapping f has a fixed point.

Example 6: Iterated solution to linear equations

To apply Banach's theorem 1.3 we require a complete metric space. Take the set X of all ordered n-tuples of real numbers $x = (x_1,x_2,\ldots,x_n)$, $y = (y_1,y_2,\ldots,y_n)$, $z = (z_1,z_2,\ldots,z_n)$, etc.

On X the norms ℓ^∞, ℓ^1 and ℓ^2 define respectively d_∞, d_1, and d_2 metrics such that the space $X = (X,d)$ is a Banach space. Define the transformation $f:X \to X$ by

$$y = fx = Ax + b$$

or

$$y_i = \sum_{r=1}^{n} a_{ir} x_r + b \tag{1.7}$$

where $A = \{a_{ij}\} \in \overline{M}_n$ (the set of constant n×n matrices) and $b = (b_1,\ldots,b_n) \in X$ is a constant vector. For the d_∞ metric and $z = fw$,

$$\begin{aligned} d_\infty(y,z) &= d_\infty(fx,fw) = \max_j |y_j - z_j| \\ &= \max_j \left| \sum_{r=1}^{n} a_{jr}(x_r - w_r) \right| \\ &\le d_\infty(x,w) \max_j \sum_{r=1}^{n} |a_{jr}| \\ &\le k\, d_\infty\, (x,w), \end{aligned} \tag{1.8}$$

where $k = \max_j \sum_{r=1}^{n} |a_{jr}|$. So if $\sum_{r=1}^{n} |a_{jr}| < 1$ for all j then $x = Ax + b$ has a unique solution x which is given by the limit of the iterated sequence $x^{(0)}, x^{(1)}, \ldots$ to

$$x^{(k+1)} = Ax^{(k)} + b, \tag{1.9}$$

for $k = 0,1,\ldots$ and $x^{(0)}$ arbitrary. The contraction condition $\sum_{r=1}^{n} |a_{jr}| < 1$ for the ℓ^∞ norm is a row sum, if instead the ℓ^1 norm was used then the contraction condition would be $\sum_{j=1}^{n} |a_{jr}| < 1$ for all r (column sum condition), and similarly if the ℓ^2 norm was used the contraction condition would be $\sum_{j=1}^{n} \sum_{k=1}^{n} |a_{jk}|^2 < 1$.

The iteration algorithm (1.9) can be used to solve the linear equation $Cx = g$, $\det C \neq 0$ by setting $C = H-G$ with $\det H \neq 0$, so that on rearrangement

$$x = H^{-1}Gx + H^{-1}g \qquad (1.10)$$

Clearly if we put $A = H^{-1}G$ and $b = H^{-1}g$ the algorithm (1.9) can be used directly to solve for x. Two iteration schemes based upon this decomposition idea are Jacobi and Gauss-Seidal iteration

Banach's fixed point theorem has many other applications including existence and uniqueness theorems for ordinary differential equations and for Fredholm and Volterra integral equations.

1.4 Linear Operators

Definition 1.16: *Linear operator*

A *linear operator* $f:X \to Y$ is such that

(i) the domain, D(f), of f is a vector space X and the range, R(f), of f lies in a vector space Y over the same field F.

(ii) For all $x,z \in D(f)$ and scalars $\alpha \in F$

$$\begin{aligned} f(x+z) &= fx + fz \\ f(\alpha x) &= \alpha f(x) \end{aligned} \qquad (1.11)$$

A generalisation of the above definition is if $\alpha_i \in F$, and $x_i \in D(f)$, for $i = 1,2,\ldots,n$, then $f(\sum_{i=1}^{n} \alpha_i x_i) = \sum_{i=1}^{n} \alpha_i f x_i$.

This property of linear operators is useful in matrix analysis and in differential equations and is called the principle of superposition.

Example 7

The following are examples of linear operators:

(i) The vector space X consisting of all polynomials on the interval [a,b]; then

$$fx(t) = \dot{x}(t) \qquad \text{for every} \quad x \in X$$

is the linear differential operator and maps X into itself.

(ii) A linear operator f from C[a,b] into itself can be defined by

$$fx(t) = \int_0^t x(s)ds.$$

(iii) The real $(n\times r)$ matrix $A = \{a_{ij}\}$ defines a linear operator $f:R^n \to R^r$ by means of the equation $y = Ax$.

If the mapping $f:D(f) \to Y = R(f)$ is one to one then there exists the linear inverse mapping $f^{-1}:R(f) \to D(f)$. Clearly $f^{-1}fx = x$ for all $x \in D(f)$ and $f^{-1}fy = y$ for all $y \in R(f)$. The inverse of a linear operator exists if and only if the null space of the operator consists only of the zero vector, that is $fx = 0 \Rightarrow x = 0$.

If we now consider linear normed vector spaces we have:

Definition 1.17: *Bounded linear operators*

Let X and Y be linear normed spaces and $f:D(f) \to Y$ be a linear operator, where $D(f) \in X$. The operator f is said to be *bounded* if there is a real number $c > 0$ such that

$$\|fx\| \leq c\|x\| \qquad \text{for all} \quad x \in D(f).$$

Clearly there is a smallest c such that the above inequality is satisfied and we call this the *induced norm* of the operator f. Corresponding to the vector norm $\|x\|$, the induced norm associated with the linear operator f, is

$$\min(c) = \|f\|_i = \sup_{\substack{x\in D(f)\\ x\neq 0}} \left\{\frac{\|fx\|}{\|x\|}\right\} \tag{1.12}$$

or equivalently

$$\|f\|_i = \sup_{\substack{x\in D(f)\\ \|x\|=1}} \|fx\| \tag{1.13}$$

the above definitions imply that

$$\|fx\| \leq \|f\|_i \, \|x\| \tag{1.14}$$

Clearly $\|f_i\|$ can be interpreted as the maximum gain of the mapping f. We note that the induced norm satisfies conditions (**n1-n4**) of definition **1.11**, and that an induced norm is topologically equivalent to another non-induced norm. A special property of induced norms for linear transformations defined by matrices is that they are submultiplicative, that is $\|AB\|_i \leq \|A\|_i \|B\|_i$ for all matrices $A,B \in R^{n\times n}$. Finally, if the space Y is a Banach space then it can be shown that the space of all bounded linear operators $f:X \to Y$ with the induced norm is also a Banach space.

Example 8

(i) In example 7(i) the differential operator $fx(t) = \dot{x}(t)$ was defined on the space X, of all polynomials t^n on the interval $I = [0,1]$ with norm $\|x\| = \max_{t\in I} |x(t)|$. We see that f is linear but not bounded, since on setting $x_n = t^n$, for $n \in N$, the norm of x is $\|x_n\| = 1$ and $\|fx_n\| = n$ so that the induced norm for the differential operator $\|f\| = n$ (which is arbitrarily large).

(ii) The integral operator $f:C[0,1] \to C[0,1]$ defined by $y = fx$ where $y(t) = \int_0^1 k(t,\tau)x(\tau)d\tau$ and k is continuous on $I\times I$, $I = [0,1]$, is both linear and bounded.

(iii) Also the operator $f:R^n \to R^r$, which is defined by the equation $y = Ax$ with $A = \{a_{ij}\}$ a $(n\times r)$ matrix, is linear and bounded.

A *functional* is a special operator in that its range lies on the real line R or complex plane C; it obviously satisfies the above conclusions concerning norms, continuity and boundedness.

A general result for finite dimensional systems with normed space X, is that every linear operator on X is bounded. Further for any linear operator f boundedness and continuity are equivalent and if f is continuous at a single point it is continuous everywhere in X. Matrices are the most important tool for studying linear operators on finite dimensional vector spaces, since the operators can always be respresented by matrices.

1.5 Linear Transformations and Matrices

Let X and Y be finite dimensional vector spaces over the same field and $f:X \to Y$ be a linear operator. Let $H = \{h_1, h_2, \ldots, h_n\}$ and $G = \{g_1, g_2, \ldots, g_r\}$ act as basis vectors for X and Y respectively, with the respective vectors in fixed order. Thus each $x \in X$ has a unique representation

$$x = \alpha_1 h_1 + \alpha_2 h_2 + \ldots + \alpha_n h_n \tag{1.15}$$

where α_i are scalars. Since f is linear, x has the image

$$y = fx = f(\sum_{k=1}^{n} \alpha_k h_k) = \sum_{k=1}^{n} \alpha_k f h_k, \tag{1.16}$$

again this representation is unique so f is uniquely determined if the images $y_k = f h_k$ of the n basis vectors $h_1, \ldots, h_n$ are prescribed. Since $y, y_k \in Y$, they have unique representations

$$y = \sum_{j=1}^{n} \beta_j g_j, \quad y_k = f h_k = \sum_{j=1}^{r} a_{jk} g_j, \tag{1.17}$$

where each g_j form a linearly independent set. Therefore equating (1.16) and (1.17) gives,

$$y = \sum_{k=1}^{n} \alpha_k \sum_{j=1}^{r} a_{jk} g_j = \sum_{j=1}^{r} (\sum_{k=1}^{n} a_{jk} \alpha_k) g_j \tag{1.18}$$

Therefore,

$$\beta_j = \sum_{k=1}^{n} a_{jk} \alpha_k \qquad j = 1, 2, \ldots, r. \tag{1.19}$$

The coefficients $\{a_{jk}\}$ form a $(r \times n)$ matrix $A_{HG} = \{a_{jk}\}$ for fixed H,G, which is uniquely determined by the linear operator f. This matrix represents the operator f with respect to the bases H,G. However the matrix A_{HG} is not unique, it depends upon the basis vectors chosen in X and Y. We shall see that the operator is *uniquely* represented by an equivalence class of similar matrices

$\{A_{HG}, A_{H'G'}\}$ for $X = Y = R^n$ which are such that $A_{HG} = C^{-1}A_{H'G'}C$ for C a (n×n) nonsingular matrix.

So consider the linear operator $f:X \to X$, on a normed space X of dimension n. Let $h = \{h_1,\ldots,h_n\}$ and $\tilde{h} = \{\tilde{h}_1,\ldots,\tilde{h}_n\}$ be two row vectors which act as different bases for the vector space X, which by definition of vector basis each h_j is a linear combination of the $\tilde{h}_k$'s and conversely, that is $\tilde{h} = hC$ or $\tilde{h}' = C'h'$ where C is a nonsingular (n×n) matrix. For every $x \in X$ there is a unique representation with respect to these bases

$$\begin{aligned} x &= hx_1 = \sum_{j=1}^{n} \alpha_j h_j \\ &= \tilde{h}x_2 = \sum_{j=1}^{n} \tilde{\alpha}_j \tilde{h}_j \end{aligned} \quad \Rightarrow x_1 = Cx_2$$

Similarly for $y = fx = hy_1 = \tilde{h}y_2 \Rightarrow y_1 = Cy_2$. Consequently if A_1 and A_2 denote the matrices which represent f with respect to h and $\tilde{h}$ respectively then

$$y_1 = A_1x_1 \quad \text{and} \quad y_2 = A_2x_2$$

hence

$$CA_2x_2 = Cy_2 = y_1 = A_1x_1 = A_1Cx_2$$

or

$$A_2 = C^{-1}A_1C, \tag{1.20}$$

which is the definition of *similar matrices* A_1, A_2. Also the *characteristic determinants* of this equality gives

$$\begin{aligned} \mathrm{Det}(A_2-\lambda I) &= \mathrm{Det}(C^{-1}A_1C - \lambda C^{-1}IC) \\ &= \mathrm{Det}(A_1 - \lambda I) \end{aligned} \tag{1.21}$$

which establishes:

Theorem 1.4

All matrices representing a linear operator $f:X \to X$ on a finite dimensional normed space X, relative to various bases for

X have the same characteristic values. Also two matrices representing the same linear operator f are similar, each with the same characteristic values.

Since the characteristic equation of the operator f is of order n, f has at least one characteristic value (and at most n) for $X \neq \{0\}$.

An important aspect of similarity transformations is the possibility of reducing a matrix to block diagonal form to ease computation. Suppose that X is a n-dimensional vector space over some field F. Then any non-zero vector $e^{(i)} \in X$ is said to be a *characteristic* vector of the linear operator $f:X \to X$ if

$$A_1 e^{(i)} = \lambda_i e^{(i)} \tag{1.22}$$

where A_1 is a (n×n) matrix associated with f and $\lambda_i \in F$ is a characteristic value of A given by solution of $\mathrm{Det}(A_1 - \lambda I) = 0$. Define the matrix $C = (e^{(1)}, \ldots, e^{(n)})$ (the so-called *modal matrix*) then from (1.22)

$$A_1 C = C\,\mathrm{Diag}(\lambda_i) \quad \text{or} \quad \mathrm{Diag}(\lambda_i) = C^{-1} A_1 C \tag{1.23}$$

so that A_1 is similar to a diagonal matrix that contains the characteristic values of A_1 along its diagonal. In this case we have assumed that the transforming matrix C has n linearly independent column vectors each associated with n distinct characteristic values of the matrix A_1. It is important to note that although similar matrices have the same characteristic values the corresponding characteristic vectors are not necessarily the same. In the more general case when A_1 has multiple characteristic values, it can be reduced via a similarity transformation to a unique block diagonal form,

$$C^{-1} A_1 C = J \tag{1.24}$$

where

$$J = \begin{pmatrix} J_1 & 0 & . & . & 0 \\ 0 & J_2 & & & . \\ . & & . & & . \\ & & & . & . \\ 0 & . & . & . & J_m \end{pmatrix} \quad \text{and} \quad J_i = \begin{pmatrix} \lambda_i & 0 & . & . & . & . & 0 \\ 1 & \lambda_i & 0 & & & & . \\ 0 & 1 & \lambda_i & & & & . \\ 0 & 0 & 1 & . & & & . \\ . & & & & . & & . \\ . & & & & & . & 0 \\ 0 & 0 & 0 & . & . & 1 & \lambda_i \end{pmatrix}$$

is a $(r_i \times r_i)$ matrix with r_i the multiplicity of the ith characteristic value of A_1 and such that $\sum_{i=1}^{m} r_i = n$. This *canonical form* is called the *Jordan* form for A_1, in this case the characteristic equation for A_1 can be written as $\prod_{i=1}^{m} (\lambda-\lambda_i)^{r_i} = 0$.

Consider matrices over the field of complex numbers, and denote the complex conjugate of matrix A by A^*. The matrix A is said to be *normal* when $AA^* = A^*A$, special cases of normal matrices are *Hermitian* matrices and *Unitary* matrices which satisfy $A^* = A$ and $AA^* = A^*A = I$ respectively. All normal matrices can be diagonalised by using unitary matrices in the similarity transformation, therefore any $(n \times n)$ normal matrix has n orthonormal characteristic vectors. A consequence of this is that all unitary matrices have characteristic values that lie on the unit circle. The location of the characteristic values of diagonal and triangular matrices clearly lie on the main diagonal. In the case of more generalised matrices characteristic value location is given by Gershgorin's theorem:

Theorem 1.5

An arbitrary $(n \times n)$ matrix $A = \{a_{ij}\}$ over the complex field has its characteristic values located in the union of circles in the complex plane z defined by

$$|z - a_{ii}| \leq r_i, \qquad i = 1,2,\ldots,n \tag{1.25}$$

where

$$r_i = \sum_{\substack{j=1\\ j\neq i}}^{n} |a_{ij}| \quad \text{or} \quad \sum_{\substack{i=1\\ i\neq j}}^{n} |a_{ij}|.$$

Clearly the more diagonally dominant A is, the smaller the Gershgorin circles and the tighter the bounds on the location of the characteristic values of A.

In the remainder of this section we introduce the concept of a measure of a matrix, $\mu(A)$; this has the advantage of allowing tight estimates to be made upon the location of characteristic values of a matrix (dependent upon the norm used) since $\mu(A)$ unlike the induced norm $\|A\|_i$ can have negative values.

Consider the complex Euclidean space $X = C^n$; from the norm condition $|\,\|x\| - \|y\|\,| \leq \|x\| - \|y\|$ it follows for any two vectors $x, y \in X$ that the limit $\mu(\|x\|) = \lim_{\alpha\to 0^+} \left\{\frac{\|x+\alpha y\| - \|x\|}{\alpha}\right\}$ exists, since the function inside the limit is a nondecreasing function of α and bounded by $\pm\|y\|$. The function $\mu : C^n \to R$ is a one-sided derivative of the norm function. If the vector norm in the definition of $\mu(\cdot)$ is replaced by the induced norm for a matrix A, we have the definition of the *measure of a matrix*,

$$\mu(A)_i = \lim_{\alpha\to 0^+} (\|I+\alpha A\|_i - 1)\alpha^{-1} \tag{1.26}$$

It is easily seen from this definition and the algebraic properties of vector norms that for any $A, B \in C^n \times C^n$ and all $\alpha \in R_+$ that

$$\mu(\alpha A)_i = \alpha\mu(A)_i$$

$$|\mu(A+B)_i| \leq \mu(A)_i + \mu(B)_i$$

$$|\mu(A)_i - \mu(B)_i| \leq \|A - B\|_i$$

$$\|A\|_i \geq \mu(A)_i \geq -\mu(-A)_i \geq -\|A\|_i \tag{1.27}$$

Also if $\lambda_j(A)$ $(j=1,2,\ldots,n)$ is a characteristic value of matrix A and if its associated normalised characteristic vector is $e^{(j)}$, then as $\alpha \to 0^+$

$$\alpha^{-1}(\|(I+\alpha A)e^{(j)}\|_i - \|e^{(j)}\|) = \alpha^{-1}(|1+\alpha\lambda_j(A))|-1) \to \mathrm{Re}(\lambda_j(A)) \tag{1.28}$$

and

$$\alpha^{-1}(\|(I+\alpha A)e^{(j)}\|_i - \|e^{(j)}\|) \le \alpha^{-1}(\|(I+\alpha A)\|_i - 1) \to \mu(A)_i \tag{1.29}$$

so that

$$\mu(A)_i \ \ge \ \mathrm{Re}(\lambda_j(A)), \qquad \text{for } i = 1,2,\ldots,n \tag{1.30}$$

The computation of the measure of a (n×n) matrix $A = \{a_{ij}\}$ is relatively straightforward; consider the space $X = C^n$ with the vector norms ℓ^1, ℓ^2 and ℓ^∞, their respective matrix induced norms are $\sup_j(\sum_{i=1}^{n} |a_{ij}|)$ (column sum), $[\max_i \lambda_i(A^*A)]^{\frac{1}{2}}$, and $\sup_i(\sum_{j=1}^{n} |a_{ij}|)$ (row sum); similarly the matrix measure $\mu(A)$ associated with these induced norms are respectively $\mu(A)_1 = \sup_j (\mathrm{Re}(a_{jj}) + \sum_{i=1, i\neq j}^{n} |a_{ij}|)$, $\mu(A)_2 = \max_i \lambda_i(\frac{1}{2}(A+A^*))$, and $\mu(A)_\infty = \sup_i(\mathrm{Re}(a_{ii}) + \sum_{j=1, j\neq i}^{n} |a_{ij}|)$.

Example 9

Consider the (2×2) real matrix $A = \begin{pmatrix} 3 & 1 \\ 0 & 2 \end{pmatrix}$, its characteristic values are 2,3, and its associated ℓ^1, ℓ^2 and ℓ^∞ matrix measures are $\mu(A)_1 = 3$, $\mu(-A)_1 = -1$, $\mu(A)_2 = \frac{5}{2} + \frac{\sqrt{2}}{2}$, $\mu(-A)_2 = -\frac{5}{2} + \frac{\sqrt{2}}{2}$, $\mu(A)_\infty = 4$, $\mu(-A)_\infty = -2$. So that the characteristic value for bounds for these respective matrix measures are given by inequality (30) as

ℓ^1: $1 \le \mathrm{Re}(\lambda) \le 3$,

ℓ^2: $\frac{5}{2} - \frac{\sqrt{2}}{2} \le \mathrm{Re}(\lambda) \le \frac{5}{2} + \frac{\sqrt{2}}{2}$,

ℓ^∞: $2 \le \mathrm{Re}(\lambda) \le 4$.

Note that the upper bound matrix measure on the ℓ^1 norm and the lower bound on the ℓ^∞ norm give the exact characteristic values of the matrix A. Whilst the matrix measure associated with the ℓ^2 norm gives the narrowest bandwidth for the location of the characteristic values of A.

The above example shows that matrix measures can have negative values and therefore cannot be a norm (unlike the induced norm for which $\|-A\|_i = \|A\|_i$), however this very property is particularly useful in estimating the bounds of solutions to ordinary differential equations.

Consider the homogeneous differential equation $\dot{x} = A(t)x$, where A(t) is a (n×n) continuous matrix defined for $t \geq t_o$; then $\|x(t)\|$ is a solution to this differential equation with right handed (*Dini*) derivative

$$\begin{aligned} D^+(\|x\|) &= \lim_{\alpha\to 0^+} (\|x+\alpha\dot{x}\| - \|x\|)\alpha^{-1} \\ &= \lim_{\alpha\to 0^+} (\|(I+\alpha A)x\| - \|x\|)\alpha^{-1} \\ &\quad \lim_{\alpha\to 0^+} \{(\|(I+\alpha A)\|_i - 1)\|x\|\}\alpha^{-1} \\ &= \mu(A)_i\|x\| \end{aligned} \tag{1.31}$$

Remembering that $\mu(A) \geq -\mu(-A)$ integrating (1.31) over $[t_o,t]$ and utilising the properties (1.27) of the matrix measure we get *Coppel's inequality*,

$$\|x(t_o)\| \exp\left\{-\int_{t_o}^{t} \mu(-A(\tau))d\tau\right\} \leq \|x(t)\| \leq \|x(t_o)\| \times \exp\left\{\int_{t_o}^{t} \mu(A(\tau))d\tau\right\} \tag{1.32}$$

From which it follows that if $x(t_o) = 0$ the null solution is the unique solution (if it exists) to $\dot{x} = A(t)x$. Further, if $x(t_o) \neq 0$, $x(t;x_o,t_o)$ is a unique solution to $\dot{x} = A(t)x$, since the difference of two solutions to $\dot{x} = A(t)x$ is by linearity also a solution (if it exists). The general existence and uniqueness conditions for solutions to differential equations are given by the contraction mapping theorem **1.3** since $\|A\|_i \geq \mu(A)_i \geq -\mu(A)_i \geq -\|A\|_i$, the measure of a matrix in Coppel's inequality can be replaced by the induced norm of the matrix, but clearly the resulting bounds on solution will not be as tight.

1.6 Inner Product Spaces and Fourier Series

The concepts of dot product and orthogonality of vector algebra can be generalised to arbitrary normed vector spaces; this leads to inner product spaces and complete inner product spaces which are called Hilbert spaces. These spaces retain many of the qualities of Euclidean spaces and geometry, in particular that of orthogonality and projections.

Definition 1.18

An *inner product space* X is a vector space with an inner product $\langle x,y\rangle$ defined on it, that is for every pair of vectors x,y there is a scalar $\langle x,y\rangle$ such that

p1. $\langle x+y,z\rangle = \langle x,z\rangle + \langle y,z\rangle$

p2. $\langle \alpha x,y\rangle = \alpha\langle x,y\rangle$

p3. $\langle x,y\rangle = \overline{\langle y,x\rangle}$

p4. $\langle x,x\rangle \geq 0$

p5. $\langle x,x\rangle = 0 \iff x = 0$

An inner product on X defines a norm $\|x\| = \langle x,x\rangle^{\frac{1}{2}}$ on X and a metric $d(x,y) = \|x-y\|$, hence an inner product spaces are normed spaces. A complete inner product space is called a *Hilbert space*, and also a Banach space. Also since the normed linear space $(X,\|\cdot\|)$ has a uniformly continuous norm $\|\cdot\|$ on X, then similarly for the inner product space $(X,\langle\cdot,\cdot\rangle^{\frac{1}{2}})$ the function $x \mapsto \langle x,y\rangle$ is uniformly continuous for each $y \in X$.

All norms on inner product spaces satisfy the additional so-called *parallelogram equality*, p6. $<x+y,x+y> + <x-y,x-y> = 2(<x,x>+<y,y>)$.

This equality does not hold for other norms and clearly not all normed spaces are inner product spaces.

Utilising the properties of norms and inner product spaces we have the following additional special properties for inner product spaces,

p7. $<\alpha x,\alpha x> = |\alpha|^2 <x,x>$

p8. $|<x,y>| \leq <x,x>^{\frac{1}{2}} <y,y>^{\frac{1}{2}}$

p9. $<x+y,x+y> \leq <x,x> + <y,y>$ (Triangle inequality)

Property p8 is Schwartz inequality which becomes an equality if $\{x,y\}$ is a linearly independent set. A final and most important concept of inner product spaces is if $x,y \in X$ are such that

p10. $<x,y> = 0$,

then vectors x and y are *orthogonal*.

Example 10

(a) The Euclidean spaces R^n and C^n are both Hilbert spaces with inner products defined by

$$<x,y> = \alpha_1\beta_1^* + \ldots \alpha_n\beta_n^*$$

for $x = (\alpha_1,\alpha_2,\ldots,\alpha_n)$ and $y = (\beta_1,\beta_2,\ldots,\beta_n)$

(b) The space $L^2[a,b]$. The vector space of all continuous valued functions on [a,b] forms a normed space X with norm

$$||x|| = <x,x>^{\frac{1}{2}} = \left\{\int_a^b x(t)^2 dt\right\}^{\frac{1}{2}}, \qquad x \in L^2[a,b].$$

Also for any $x,y \in L^2[a,b]$ which are complex valued, their inner product is defined by

$$<x,y> = \int_a^b x(t)y(t)^* dt$$

and norm is given by

$$||x|| = \left\{\int_a^b |x(t)|^2 dt\right\}^{\frac{1}{2}}.$$

But since $x(t)x(t)^* = |x(t)|^2$, then $L^2[a,b]$ is a Hilbert space. We also note that for any $x,y \in L^2[a,b]$ Holder's inequality becomes Schwartz inequality and Minkowski's inequality becomes the triangle inequality.

(c) The sequence space ℓ^2 with inner product $\langle x,y\rangle = \sum_i^\infty \alpha_i \beta_i^*$ is a Hilbert space, but ℓ^p $(p \neq 2)$ is *not* an inner product space nor a Hilbert space (it is however a Banach space since ℓ^p is complete).

(d) Similarly the function space C[a,b] is a Banach space but not an inner product space nor a Hilbert space.

Consider now the Euclidean space R^3 with basis vectors h_1, h_2, h_3 then any $x \varepsilon R^3$ has the unique representation

$$x = \alpha_1 h_1 + \alpha_2 h_2 + \alpha_3 h_3$$

Taking inner products with h_1, h_2, and h_3 enables us to calculate the unknown coefficients $\alpha_1, \alpha_2, \alpha_3$ as

$$\langle x,h_1\rangle = \alpha_1\langle h_1,h_2\rangle + \alpha_2\langle h_2,h_1\rangle + \alpha_3\langle h_3,h_1\rangle = \alpha_1 k_1$$

where $k_1 = \langle h_1,h_1\rangle$, and $\langle h_2,h_1\rangle = \langle h_3,h_1\rangle = 0$ by the orthogonality property. If $k_i = 1$ then we say that the basis set $\{h_i\}$ is *orthonormal* otherwise it is *orthogonal*. In general for $x \varepsilon R^n$ and $\{h_k\}$ an orthonormal sequence in an inner product space X, we have the unique representation

$$x = \sum_{k=1}^{n} \langle x,h_k\rangle h_k \tag{1.33}$$

where the coefficients $\langle x,h_k\rangle$ are independent of n. Clearly an orthonormal set $\{h_j\}$ is linearly independent, conversely any arbitrary linearly independent $\{g_k\}$ in $(X,\langle\cdot,\cdot\rangle)$ can be orthonormalised into a sequence $\{h_j\}$ in $(X,\langle\cdot,\cdot\rangle)$ by noting that $\text{span}(h_1, \ldots, h_n) = \text{span}(g_1,\ldots,g_n)$; the resulting procedure (Gram-Schmidt process) gives the relationship between the sequences $\{h_k\}$ and $\{g_k\}$ as

$$h_k = v_k \, \|v_k\|^{-1} \tag{1.34}$$

where

$$v_k = g_k - \sum_{r=1}^{k-1} \langle g_k, h_r \rangle h_r$$

Example 11

Consider the inner product space $(X, \langle\cdot,\cdot\rangle)$ for all real valued continuous functions on $[\pi,-\pi]$ with inner product $\langle x,y\rangle = \int_{-\pi}^{\pi} x(t)y(t)dt$.

An orthogonal sequence in X is

$$u_n = \cos nt , \quad n = 0,1,2,...$$

or

$$v_n = \sin nt , \quad n = 1,2,...$$

$$\text{since} \quad \langle u_m, u_n \rangle = \int_{-\pi}^{\pi} \cos mt \cos nt \, dt = \begin{cases} 0 & m \neq n \\ \pi & m=n=1,2,... \\ 2\pi & m=n=0 \end{cases}$$

Hence an orthonormal sequence in X is,

$$\tilde{u}_m = u_m \|u_m\|^{-1} = \frac{\cos mt}{\sqrt{\pi}} \quad \text{for} \quad m = 1,2,...; \quad \tilde{u}_o = \frac{1}{\sqrt{2\pi}}$$

Similarly for v_n, its orthogonal sequence in X is

$$v_n = \frac{\sin nt}{\sqrt{\pi}} , \quad n = 1,2,...$$

Returning to the unique representation (1.33) of $x \in R^n$ with respect to the orthonormal sequence h_k , the coefficients $\langle x,h_k\rangle$ are called *Fourier coefficients*. By taking norms of expression (1.33) and utilizing the Pythagorean relation then for every $x \in X$

$$\sum_{k=1}^{n} |\langle x,h_k\rangle|^2 \leq \|x\|^2 , \qquad (1.35)$$

which is known as *Bessels inequality*. Also it is not difficult to show that for the space $(X, \langle\cdot,\cdot\rangle)$ and any $x,y \in X$ that,

$$\sum_{k=1}^{n} |\langle x,h_k\rangle\langle y,h_k\rangle| \leq \|x\|.\|y\| \tag{1.36}$$

Clearly Bessel's inequality is a special case of the above inequality.

In the remainder of this section we take $X = L^2$. Let $N \subset X$ be the linear space spanned by an orthonormal sequence $\{h_1,\ldots, h_n\ldots\}$ in L^2 and if we consider the smallest subset $\overline{N}$ (the closure of N) of L^2 containing N, then if $f \in L^2$ the *Fourier series* of f converges in the mean (L^2 norm) to the orthogonal projection g of f on $\overline{N}$ ($g \in \overline{N}$); in particular the Fourier series of f converges in the mean to f if and only if $f \in \overline{N}$. If in addition N is dense in L^2 (more generally in the Hilbert space $(X,\langle\cdot,\cdot\rangle)$), that is $\overline{N} = L^2$ (more generally $\overline{N} = X$) we say that the orthonormal set $\{h_j\}$ is *complete* or *total* in L^2 (respectively in $(X,\langle\cdot,\cdot\rangle)$), and for any $f \in L^2$ ($f \in (X,\langle\cdot,\cdot\rangle)$) the Fourier series of f with respect to $\{h_j\}$ converges in the mean (in the norm $\langle\cdot,\cdot\rangle$) to f. That is the partial sum

$$x_n = \alpha_1 h_1 + \ldots + \alpha_n h_n \;; \quad \alpha_k = \langle x,h_k\rangle$$

converges to x in the mean, or $\lim_{n\to\infty} \|x_n - x\| = 0$. In example *11* the sequences $\tilde{v}_m$ and $\tilde{u}_m$ form a complete orthonormal sequence. For complete orthonormal sequences Bessels inequality (1.35) becomes *Parseval's equation*,

$$\sum_k |\langle x,h_k\rangle^2| = \|x\|^2 \tag{1.37}$$

and inequality (1.36) becomes

$$\sum_k |\langle x,h_k\rangle\langle y,h_k\rangle| = \|x\| \; \|y\| \tag{1.38}$$

for $x,y \in L^2$ (or $x,y \in (X,\langle\cdot,\cdot\rangle)$). Parseval's equation is a necessary and sufficient condition for convergence in the norm of the generalised Fourier series (1.33).

We know that the classical Fourier series for a real valued function $x:R \to R$ which is periodic with period 2π (i.e. $x \in$

$L^1[-\pi,\pi]$) is

$$x(t) = \tfrac{1}{2}a_o + \sum_{k=1}^{\infty} (a_k \cos kt + b_k \sin kt) \tag{1.39}$$

with $a_k = \frac{1}{\pi}\int_{-\pi}^{\pi} x \cos kt\, dt$, $b_k = \frac{1}{\pi}\int_{-\pi}^{\pi} x \sin kt\, dt$, which both exist since they are dominated by $|x|$. By setting $u_k = \cos kt$, $v_k = \sin kt$ in the above definitions of a_k, b_k, multiplying by u_j (and respectively by v_j) and integrating with respect to t over $[-\pi,\pi]$ we obtain

$$a_j = \langle x,\tilde{u}_j\rangle \|u_j\|^{-1}, \quad b_j = \langle x,\tilde{v}_j\rangle \|v_j\|^{-1} \tag{1.40}$$

where $\tilde{u}_j$, $\tilde{v}_j$ are defined in example *11*. So that the Fourier series (1.39) can now be rewritten as

$$x(t) = \langle x,\tilde{u}_o\rangle\tilde{u}_o + \sum_{k=1}^{\infty} \{\langle x,\tilde{u}_k\rangle\tilde{u}_k + \langle x,\tilde{v}_k\rangle\tilde{v}_k\} \tag{1.41}$$

which justifies our earlier terminology of Fourier coefficients for $\langle x,h_k\rangle$.

Consider now the partial sum S_n of the first (n+1) terms of the Fourier series (1.39), as

$$S_n(t) = \tfrac{1}{2}a_o + \sum_{k=1}^{n} (a_k \cos kt + b_k \sin kt)$$

$$= \frac{1}{\pi}\int_{-\pi}^{\pi} x(\tau)\left\{\tfrac{1}{2} + \sum_{k=1}^{n} (\cos k\tau \cos k\tau + \sin k\tau \sin k\tau)d\tau\right\}$$

$$= \frac{1}{\pi}\int_{-\pi}^{\pi} x(\tau)D_n(\tau - t)d\tau \tag{1.42}$$

$$= \frac{1}{\pi}\int_{-\pi}^{\pi} \{x(t+\tau) + x(t-\tau)\}D_n(\tau)d\tau \tag{1.43}$$

where $D_n(\tau) = \frac{\sin(n+\frac{1}{2})}{2\sin\tau/2}$ for $\tau \neq 2\pi r$, r any integer, is an even function and is called the *Dirichlet kernel*. Setting $x(t) = 1$ in (1.42) gives $1 = \frac{1}{\pi}\int_0^{\pi} 2D_n(t)dt$, multiplying this result by x(t) and subtracting from (1.42) gives

$$\frac{1}{\pi}\int_0^{\pi} \{x(t+\tau) + x(t-\tau) - x(t)\}D_n(\tau)d\tau = S_n(t) - x(t) \quad (1.44)$$

so that $S_n(t) \to x(t)$ pointwise if the above integral tends to zero as $n \to \infty$. This sufficient condition for convergence is called *Dini's condition* and is certainly satisfied if x(t) is differentiable. Surprisingly the same result on the convergence of $\{S_n\}$ can be governed by an arbitrary small interval $[-\alpha\pi,\alpha\pi]$ of $[-\pi,\pi]$, for $0 < \alpha \leq 1$, in spite of the fact that the Fourier series depends upon the whole of the interval $[-\pi,\pi]$. If in addition x is of bounded variation on the restricted interval $[-\alpha\pi, \alpha\pi]$ then by considering the upper bounds of $\int D_n(\tau)d\tau$, the Fourier series of x at a discontinuity at points t converges to $\frac{1}{2}\{x(t+0) + x(t-0)\}$, this is known as *Jordans condition*.

The fact that the Fourier series of continuous functions need not converge everywhere endangered the whole theory of representation of a function by its Fourier series. This situation was salvaged by Fejer who showed that the Fourier series of a continuous function x(t) is *summable* to x(t) by the method of *arithmetic means* (or *Cesaro sums*). Essentially this averaging process smooths out the oscillations caused by the method of partial sums which utilise Dirichlet kernels. Let

$$\sigma_n(t) = (S_0(t) + \ldots + S_n(t))(n+1)^{-1} \quad \text{for } n=0,1,2,.. \quad (1.45)$$

which is clearly the arithmetic mean of the first n partial sums $S_n(t)$ of the continuous function x(t). Also from equation (1.42)

$$\sigma_n(t) = \frac{1}{(n+1)\pi}\int_{-\pi}^{\pi} x(\tau)F_n(\tau-t)d\tau \quad (1.46)$$

where $F_n(t) = (D_o(t) + \ldots + D_n(t)) = \sin^2\frac{(n+1)t}{2}\,(2\sin^2\frac{t}{2})^{-1} > 0$, if $2n\pi \neq t$. The kernels $F_n(t)$ are the well known *Fejer kernels* which converge positively to zero as $n \to \infty$. Since the Fejer kernels are positive, of period 2π and satisfy the same integral type equations as the Dirichlet kernels then it is not difficult to see that if $S_n(t) \to x(t)$ then also $\sigma_n(t) \to x(t)$ pointwise as $n \to \infty$. Fejer's important result, that parallels that of Jordons condition, is that if $x \in L^1[-\pi,\pi]$, then for any discontinuity t at which the limits x(t+0), x(t-0) exist the Fourier series of x(t) is Cesaro summable to $\frac{1}{2}\{x(t+0)+x(t-0)\}$. Essentially this result shows that the Fourier series of a continuous function x of period 2π is Cesaro summable at *every* point t to the function, in addition the series $\{\sigma_n\}$ converges uniformly to x. Also since the power series for the sine and cosine are uniformly convergent on a bounded interval, the Fejer approach can be used to approximate any real valued continuous function x(t) on a bounded closed interval [a,b] by a polynomial p(t) such that $|x(t)-p(t)| < \varepsilon$ for any $\varepsilon > 0$ and for all $t \in [a,b]$ - this is the classical *Weierstrass approximation theory*.

1.7 Notes

Throughout this chapter results mainly germane to the remainder of this text have been presented without formal proof. The style and approach adopted is similar to that of Curtain and Pritchard (1977) which appears in the same series. A suitable introductory text in functional analysis is that of Naylor and Sell (1971), whilst a more advanced text by Bachman and Narici (1966) provides the majority of the omitted proofs of this chapter. The classical text of Dunford and Schwartz (1963) provides the necessary background in topology and linear operators. Results on linear transformations and linear algebra can be found in the very readable text of Hadley (1961), whilst Luenberger

(1969) provides an excellent introduction to vector space methods for readers with an engineering mathematics background. The results on inner product spaces and Hilbert spaces can be found in the specialist text of Halmos (1957). Finally, the material on matrix induced norms and measure of a matrix is summarised in Coppel's (1965) text on stability theory.

References

Bachman, G. and Narici, L. (1966). "Functional analysis", Academic Press, New York

Coppel, W.A. (1965). "Stability and asymptotic behaviour of differential equations", Heath, Boston

Curtain, R.F. and Pritchard, A.J. (1977). "Functional analysis in modern applied mathematics", Academic Press, New York

Dunford, N. and Schwartz, J. (1963). "Linear operators", Vols.I, II, J. Wiley, Interscience, New York

Hadley, G. (1961). "Linear algebra", Addison Wesley, New York

Halmos, P. (1957). "Introduction to Hilbert spaces", Chelsea, New York

Luenberger, D.G. (1969). "Optimization by vector space methods", J. Wiley, New York

Naylor, A.W. and Sell, G.R. (1971). "Linear operators in engineering and science", Holt, Rinehart and Winson, New York

Chapter 2

ALMOST PERIODIC FUNCTIONS

2.1 Introduction

The theory of almost periodic functions was created and developed in its main features by Bohr (1924) as a generalisation of pure periodicity. The general property can be illustrated by means of the particular example

$$f(t) = \sin 2\pi t + \sin 2\pi t\sqrt{2} .$$

This continuous function is not periodic: that is there exists no value of τ which satisfies the equation $f(t+\tau) = f(t)$ for all values of t. However, we can establish the existence of numbers for which this equation is approximately satisfied with an arbitrary degree of accuracy. For given any $\eta > 0$ as small as we please we can always find an integer τ such that $\tau\sqrt{2}$ differs from another integer by less than $\eta/2\pi$. It can be shown that there exist infinitely many such numbers τ, and that the difference between two consecutive numbers is bounded. This property of the τ's defines almost periodicity in general (Fink, 1974).

Almost periodicity is a structural property of functions which is invariant with respect to the operations of addition and multiplication, and also in some cases with respect to divison, differentiation, integration and other limiting processes. To the structural affinity between almost periodic functions and purely periodic functions may be added an analytical similarity. To any almost periodic function there corresponds a Fourier series in the form of a general trigonometric series

$$f(t) \simeq \sum_{k=1}^{\infty} A_k \exp(i\lambda_k t)$$

with λ_k being real numbers and A_k real or complex. The series is obtained from the function by the same formal process as in the case of purely periodic functions, that is, by the method of undetermined coefficients and term by term integration. As in the purely periodic case, the Fourier series need not converge to the almost periodic function for all values of t. Nevertheless, there is still a very close connection between the series and the function. In the first place Parseval's equation holds;

$$M_t\{|f(t)|^2\} = \sum_{k=1}^{\infty} |A_k|^2,$$

where the mean value M_t is defined by

$$M_t\{g\} \triangleq \lim_{T\to\infty} \frac{1}{T}\int_0^T g(t)dt.$$

The uniqueness theorem, according to which there exists at most one almost periodic function having a given trigonometric series for its Fourier series follow from Parseval's theorem. Further, the series is summable to f(t) in the sense that there exists a sequence of polynomials

$$\sum_{k=1}^{\infty} d_k^{(m)} A_k \exp(i\lambda_k t) \qquad (m = 1,2,\ldots.)$$

where $d_k \in [0,1]$ and only a finite number of the d_k differ from zero for each m, such that

(i) the sequence of polynomials converges to f(t) uniformly in t;

(ii) the sequence of polynomials converges to the Fourier series associated with f(t), by which is meant that for each k, $d_k^{(m)} \to 1$ as $m \to \infty$.

Conversely, any trigonometric polynomial is an almost periodic function, and so is the uniform limit of a sequence of trigonometric polynomials. It is easily proved that the Fourier series of such a limit function is the formal limit of the sequence of trigonometric polynomials (Cordeneanu, 1968). Thus the class of Fourier series of almost periodic functions consists of all trigonometric series of the general type $\Sigma\, A_k \exp(i\lambda_k t)$ to which there corresponds a uniformly convergent sequence of polynomials of the type $\Sigma\, d_k^{(m)} A_k \exp(i\lambda_k t)$, $(m = 1,2,\ldots.)$ which formally converge to the series.

The first investigations of trigonometric series, other than purely periodic series, were carried out by P. Bohl (1906). He considered the class of functions represented by series of the form

$$\sum_{(k)} A_{k_1 k_2 k_3 \ldots k_\ell} \exp\left\{(k_1\omega_1 + k_2\omega_2 + \ldots + k_\ell\omega_\ell)it\right\}$$

where $\omega_1, \omega_2, \ldots, \omega_\ell$ are arbitrary real numbers and $A_{k_1 k_2 \ldots k_\ell}$ are real or complex numbers. The theory of these functions, however, follows in a more or less natural way from existing theories on purely periodic functions, rather than the theory generated by Bohr. A quite new way of studying trigonometric series is opened up by Bohr's theory of almost periodic functions and in the sequel we will use the strong correspondence between the two to develop methods of studying differential equations with almost periodic coefficients.

Our two main aims in this chapter are to review the development of the Fourier series theory of almost periodic functions and to consider the question of approximating almost periodic functions by trigonometric polynomials. Material is taken from five main sources, namely Bohr *(op.cit.)*, Bochner (1927), Besicovich (1932), Corduneanu (1968), and Fink (1974). Proofs of standard theorems are given only when some clarification is necessary, otherwise they are omitted for brevity. The final section is devoted to almost periodic functions depending uniformly on a parameter. This section is a prelude to the material contained in Chapter Six on almost periodic differential equations dependent upon a parameter. Section 2.6 briefly considers limiting cases of almost periodic functions.

2.2 Definitions and Elementary Properties of Almost Periodic Functions

Bohr developed the theory of almost periodic functions in direct analogy with the theory of purely periodic functions, althoug theorems which are decidedly trivial for purely periodic functions are no longer trivial for almost periodic functions. Theorems for purely periodic functions are trivial because the investigation of such functions can be restricted to a finite interval, say the period itself. For the almost periodic case, similar theorems can be deduced by effectively restricting interest to a finite interval called the *inclusion interval*. The numbers, corresponding to periods in the purely periodic case, which characterise the almost periodicity of an almost periodic function, are chosen from the inclusion interval and are called almost periods. We clarify the situation by considering the class of continuous functions which have the following property: for every $\eta > 0$ there exists a *translation number* $\tau(\eta) = \tau$ of $f(t)$ such that

$$|f(t+\tau) - f(t)| < \eta \qquad \text{for all } t \tag{2.1}$$

Note that the numbers $\tau(\eta)$ are arbitrarily large and this is not, however, a satisfactory situation. Examples may be constructed to show that the class of functions we have chosen does not even remain invariant under the operation of addition. Clearly something more must be said about the $\tau(\eta)$ to produce a class of functions which are well behaved. To this end Bohr introduced the concept of relative density.

Definition 2.1: *Relatively dense set (Bohr, op.cit.)*

A set T of real numbers is said to be *relatively dense* if there exists a number $\ell > 0$ such that any interval of length ℓ

contains at least one member of T. Any such number is called an *inclusion interval* of the set T.

The number $\tau(\eta)$ is called an η-translation number of a function f(t) (Bochner, 1927), and we denote the set of all translation numbers η of f(t) by $T(\eta,f(t))$. It is easily verified that the following properties hold:

(i) $T(\eta',f(t)) \supset T(\eta,f(t))$ for any $\eta' > \eta$.

(ii) If η is an η-translation number then so is $-\eta$.

(iii) If η_1, η_2 are (η_1 and η_2)-translation numbers respectively, then $\eta_1 \pm \eta_2$ is an $(\eta_1 \pm \eta_2)$-translation number.

Thus we are led to the Bohr definition of an almost periodic function f(t):

Definition 2.2: *Bohr almost periodic function (Bohr, op.cit.)*

A continuous $f:R \to E$ is called *Bohr almost periodic* if for any $\eta > 0$ the set $T(\eta,f(t))$ is relatively dense.

Henceforth let ℓ_η denote an inclusion interval of $T(\eta,f(t))$. Each $\tau \in T(\eta,f(t))$ is now called an η-almost period of f(t) and it is clear that as $\eta \to 0$, the set $T(\eta,f(t))$ becomes rarefied, whereas, in general, $\ell_\eta \to +\infty$. From the definition it follows that any continuous purely periodic function f(t) is Bohr almost periodic, since for any η the set $T(\eta,f(t))$ contains all numbers $k\omega$ (ω a period of f(t) and k an integer) and thus is relatively dense. In the sequel we shall be concerned mainly with uniformly almost periodic functions, although generalisations do exist (see section 2.6) and choose to drop the qualifier "uniformly": almost periodic will imply uniformly almost periodic unless otherwise specified. Several theorems now follow, which establish the elementary properties of almost periodic functions.

Theorem 2.1: *(Bohr, op.cit.)*

Let $f:R \to E$ be an almost periodic function, then f(t) is bounded on E.

Proof: Put $\eta = 1$ and denote by M the maximum of $|f(t)|$ in an interval $[0,\ell_1]$. It can be easily seen that corresponding to any t we can define a number $\tau \in T(1,f(t))$, such that $t + \tau$ belongs to $[0,\ell_1]$, and consequently that

$$|f(t+\tau)| < M$$

but

$$|f(t+\tau)-f(t)| < 1$$

thus

$$|f(t)| < M + 1 \quad \text{for all values of } t,$$

which proves the theorem.

The next theorem is an important statement about the continuity of almost periodic functions.

Theorem 2.2: *(Bohr, op.cit.)*

Let $f:R \to E$ be an almost periodic function then $f(t)$ is uniformly continuous on R.

Proof: Given an $\eta > 0$, take an $\ell_{\eta/3}$ and let $\delta(0<\delta<1)$ be a number such that $|f(t_1)-f(t_2)| < \frac{1}{3}\eta$ for any $t_1,t_2 \in [0,\ell_{\eta/3}+1]$ if only $|t_1-t_2| < \delta$. Let now t',t'' be any two numbers such that $|t'-t''| < \delta$. There exists a number τ of the set $T(\frac{1}{3}\eta,f(t))$ such that both the numbers $t'+\tau$, $t''+\tau$ belong to the interval $[0,\ell_{\eta/3}+1]$. We have then

$$|f(t'+\tau)-f(t''+\tau)| < \tfrac{1}{3}\eta.$$

Also

$$|f(t+\tau)-f(t)| < \tfrac{1}{3}\eta \qquad \text{for any } t.$$

Thus

$$|f(t')-f(t'')| \le |f(t')-f(t'+\tau)| + |f(t'+\tau)-f(t''+\tau)| + |f(t''+\tau)-f(t'')| < \eta$$

which proves the theorem.

The above proof exemplifies the idea of almost periodic function theory that everything that we need to know about almost periodic functions happens on some finite interval.

The following theorems **2.3-2.6** establish the algebraic and topological properties of the Bohr definition of almost periodicity.

Theorem 2.3: *(Bohr, op.cit.)*

If the complex valued function $f(t)$ is almost periodic then for any constant complex number α and real number β, $\alpha f(t)$, $f^*(t)$, $f(t+\beta)$ and $f(\beta t)$ are also almost periodic. Similarly $f(t)^2$ and $f(t)f^*(t)$ are almost periodic.

The next theorem tells something of the convergence properties of a sequence of almost periodic functions and provides the key to many results obtained in the theory of almost periodic functions, particularly with respect to polynomial approximation.

Theorem 2.4: *(Bohr, op.cit.)*

If a sequence of almost periodic functions $\{f_k(t)\}$ converges uniformly on R to a function $f(t)$, then $f(t)$ is also almost periodic.

Proof: Given η, there exists a function $f_{k_0}(t)$ such that

$$|f(t) - f_{k_0}(t)| < \frac{\eta}{3}$$

for all values of t. Let now τ be a number of $T(\frac{1}{3}\eta, f(t))$. Then

$$|f(t+\tau) - f(t)| \leq |f(t+\tau) - f_{k_0}(t+\tau)| + |f_{k_0}(t+\tau) - f_{k_0}(t)| + |f_{k_0}(t) - f(t)| < \eta$$

which shows that

$$T(\eta, f(t)) \supset T(\tfrac{1}{3}\eta, f_{k_0}(t))$$

and thus $T(\eta, f(t))$ is relatively dense. This being true for any η we conclude that $f(t)$ is almost periodic.

The following lemma tells us something about the properties of almost periods of almost periodic functions and from this we can investigate the arithmetic properties of the functions themselves.

Lemma 2.1: *(Bohr, op.cit.)*

For any $\eta > 0$ and for any two almost periodic functions $f_1(t)$ and $f_2(t)$ the set $T(\eta, f_1(t)) \cap T(\eta, f_2(t))$ is relatively dense.

Now the arithmetic properties can be deduced.

Theorem 2.5: *(Bohr, op.cit.)*

The sum of two almost periodic functions $f_1(t)$ and $f_2(t)$ is an almost periodic function.

Proof: Taking an arbitrary $\eta > 0$, let τ be any number of the set

$$T(\tfrac{1}{2}\eta, f_1(t)) \quad \cap \quad T(\tfrac{1}{2}\eta, f_2(t))$$

then

$$|f_1(t+\tau) + f_2(t+\tau) - f_1(t) - f_2(t)| \quad < \quad \eta$$

which shows that τ belongs to the set $T(\eta, f_1(t) + f_2(t))$. Thus

$$T(\eta, f_1(t) + f_2(t)) \quad \supset \quad T(\tfrac{1}{2}\eta, f_1(t)) \quad \cap \quad T(\tfrac{1}{2}\eta, f_2(t)),$$

so that $T(\eta, f_1(t) + f_2(t))$ is relatively dense, which proves the theorem.

The theorem can be generalised immediately to the case of the sum of any finite number of almost periodic functions. A similar result holds for subtraction. The next theorem is almost trivial.

Theorem 2.6

The product of two almost periodic functions $f_1(t)$ and $f_2(t)$ is an almost periodic function.

Proof: Using Theorems 2.5 and 2.3 together with the relation

$$f_1(t)f_2(t) \quad = \quad \tfrac{1}{4}\{f_1(t) + f_2(t)\}^2 - \tfrac{1}{4}\{f_1(t) - f_2(t)\}^2$$

proves the theorem.

The corresponding result for the ratio of two almost periodic functions $f_1(t)$ and $f_2(t)$ requires that $\inf_t |f_2(t)|$ is positive, since we write the ratio $f_1(t)/f_2(t)$ as the product of two almost periodic functions $f_1(t)$ and $\frac{1}{f_2}(t)$. The last two elementary properties of almost periodic functions presented concern the almost periodicity of the derivative and integral of an almost periodic function. The result for the derivative is quite straightforward, while that for the integral is a little more complex and will receive considerably more attention in later chapters in connection with the solution of differential equations

Theorem 2.7: *(Bohr, op. cit.)*

If the derivative of an almost periodic function is uniformly continuous, then it is almost periodic.

Theorem 2.8: *(Bohr, op. cit.)*

If an indefinite integral of an almost periodic function f(t) is bounded, then it is almost periodic.

Proof: is postponed until Section two, Chapter Six. A proof based on establishing the relative denseness of a set of almost periods for the integral (bounded) is rather involved and no useful purpose will be served by reproducing such a proof here. Much simpler proofs are developed from the Fourier series theory of almost periodic functions and this approach is used in Chapter Six. Theorem 2.8 is essentially a result about the solution of the differential equation $\dot{x} = f(t)$, $f(t) \sim$ almost periodic.

In the next section we examine the Fourier series theory of almost periodic functions which is developed analogously to the corresponding theory for the purely periodic case.

2.3 Mean Values of Almost Periodic Functions and their Fourier Series

We begin by recalling some of the main aspects of the Fourier series theory of periodic functions. Let f(t) be periodic with period ω, then there is a formal relation

$$f(t) \approx \sum_{|k|<\infty} A_k \exp(i\lambda_k t)$$

where

$$\lambda_k = \frac{2k\pi}{\omega} \qquad \text{and} \qquad A_k \in \overline{M}_1 \quad \text{is given by}$$

$$A_k = \frac{1}{\omega}\int_0^{\omega} f(t)\exp(-i\lambda_k t)dt \tag{2.2}$$

Questions of convergence aside, the main results are:

(i) the mapping from f(t) to the numbers $\{A_k, \lambda_k\}$ is one-to-one,

(ii) Parseval's equation holds, that is

$$\frac{1}{\omega}\int_0^{\omega} |f(t)|^2 dt = \sum_{|k|<\infty} |A_k|^2,$$

(iii) there are altered partial sums which approximate f(t).

To extend the above results to almost periodic functions we need to replace λ_k by a more general real number and replace the formula (2.2) by something more appropriate. No fixed ω will give a satisfactory theory since many almost periodic functions would have the same Fourier series.

On the other hand, the Fourier transform would not exist in general. Bohr introduced a compromise and took an average of the Fourier transform, that is, applied a limiting process to (2.2) as $\omega \to \infty$.

$$A(f,\Lambda) = \lim_{\omega\to\infty} \frac{1}{\omega}\int_0^{\omega} f(t) \exp(-i\Lambda t)dt \tag{2.3}$$

for $f(t) \varepsilon AP(C)$, where the space $AP(C) = \{f:f$ is an almost periodic complex valued function of the real variable $t\}$ with the uniform norm $||f|| = \sup_t |f(t)|$. Note that $A(f,\Lambda) \varepsilon R$ for each fixed (f,Λ). Here we have used the notation $\Lambda = \{\lambda/\lambda$ is a real number$\}$. The next important question concerns the existence of the Bohr transforms of any almost periodic function and this is answered as follows:

Theorem 2.9: *(Bohr, op.cit.)*

For any $\lambda \varepsilon \Lambda$, $A(f,\lambda)$ exists.

Proof: We first prove the theorem for $\lambda = 0$ and denote $A(f,0)$ by $M_t\{f\}$. Let ω,η be two positive numbers and m a positive integer. Write

$$\frac{1}{m\omega}\int_0^{m\omega} f(t)dt = \sum_{k=0}^{m-1} \frac{1}{m\omega}\int_{k\omega}^{(k+1)\omega} f(t)dt$$

Denote as usual by ℓ_η an inclusion interval of $T(\eta,f(t))$, and let τ_k be a number of $T(\eta,f(t))$ included in the interval $(k\omega,k\omega+\ell_\eta)$.

Then,

$$\int_{k\omega}^{(k+1)\omega} f(t)dt = \int_{k\omega-\tau_k}^{(k+1)\omega-\tau_k} f(t+\tau_k)dt$$

$$= \int_0^{\omega} f(t)dt + \int_0^{\omega} [f(t+\tau_k) - f(t)]dt$$

$$+ \int_{k\omega-\tau_k}^{0} f(t+\tau_k)dt + \int_{\omega}^{(k+1)\omega-\tau_k} f(t+\tau_k)dt$$

$$= I_1 + I_2 + I_3 + I_4 \tag{2.4}$$

Evidently $|I_2| < \eta\omega$. Now writing $M = \|f(t)\|$ and observing that the length of the range of integration in I_3 and I_4 is less than ℓ_η, we have $|I_3| < M\ell_\eta$, $|I_4| < M\ell_\eta$. Hence

$$\int_{k\omega}^{(k+1)\omega} f(t)dt = \int_0^{\omega} f(t)dt + \theta(\eta\omega+2M\ell_\eta), \tag{2.5}$$

with $|\theta| \leq 1$. Therefore

$$\frac{1}{m\omega}\int_0^{m\omega} f(t)dt = \frac{1}{\omega}\int_0^{\omega} f(t)dt + \theta(\eta + \frac{2M\ell_\eta}{\omega}), \tag{2.6}$$

where θ has changed its value but is still less than 1 in absolute value. Now let ξ be a positive number as small as we please. In the above formula set

$$\eta < \frac{\xi}{8}, \qquad \omega > \frac{16M\ell_\eta}{\xi},$$

and this gives

$$\frac{1}{m\omega}\int_0^{m\omega} f(t)dt = \frac{1}{\omega}\int_0^{\omega} f(t)dt + \frac{\theta\xi}{4} \tag{2.7}$$

Corresponding to any positive number T, define the integer m by the condition $m\omega \le T \le (m+1)\omega$. From the boundedness of f(t) we conclude that

$$\lim_{T\to\infty}\left\{\frac{1}{T}\int_0^T f(t)dt - \frac{1}{m\omega}\int_0^m f(t)dt\right\} = 0.$$

Consequently there exists a number $T_o > 0$ such that

$$\left|\frac{1}{T}\int_0^T f(t)dt - \frac{1}{m\omega}\int_0^m f(t)dt\right| < \frac{\xi}{4},$$

for all $T > T_o$. This argument shows that the limit of $\frac{1}{T}\int_0^T f(t)dt$ exists as $T \to \infty$, and if we define

$$M_t\{f\} = \lim_{T\to\infty}\frac{1}{T}\int_0^T f(t)dt \tag{2.8}$$

the first part of the theorem is proved. To solve the case of the Bohr transform, we need only consider the function f(t)exp $(-i\lambda t)$ which, for real λ, is the product of two almost periodic functions and according to Theorem 2.6 is therefore almost periodic. Thus the mean value $M_t\{f(t)\exp(-i\lambda t)\}$ exists and is defined to be $A(f,\lambda)$. This proves the assertion.

It is worth noting at this stage that the mean value M_t of almost periodic functions $f_1(t)$, $f_2(t)$ has several simple algebraic properties:-

(i) $M_t(f^*(t)) = M_t^*(f(t))$

(ii) $M_t(f(t)) \ge 0$ if $f(t) \ge 0$

(iii) $M_t(f_1+f_2) = M_t(f_1) + M_t(f_2)$

(iv) If $\{f_n(t)\}$ is a uniform convergent sequence of almost periodic functions such that $\lim_{n\to\infty} f_n(t) = f(t)$, $f(t) \in AP(C)$, then $\lim_{n\to\infty} M_t(f_n(t)) = M_t(f(t))$.

Following Bohr's development of the theory of almost periodic functions by analogy with the purely periodic case, we next look at Bessel's inequality as a first step in getting Parseval's equation. This is important because it is used to prove uniqueness for the Fourier series of an almost periodic function. The first result of interest concerns polynomial approximation to almost periodic functions.

Theorem 2.10: *(Bohr, op.cit.)*

Let f(t) be an almost periodic function; $\lambda_1, \lambda_2, \ldots, \lambda_m$ be m distinct arbitrary real numbers and $B_1, B_2, \ldots, B_m$ be m arbitrary real or complex numbers. Then

$$M_t\{|f(t) - \sum_{k=1}^{m} B_k \exp(i\lambda_k t)|^2\}$$

$$= M_t\{|f(t)|^2\} - \sum_{k=1}^{m} |A(f,\lambda_k)|^2 + \sum_{k=1}^{m} |B_k - A(f,\lambda_k)|^2 \qquad (2.9)$$

with $\qquad A(f,\lambda_k) = M_t\{f(t) \exp(-i\lambda_k t)\}.$

Proof: Write

$$M_t\{|f(t) - \sum_{k=1}^{m} B_k \exp(i\lambda_k t)|^2\}$$

$$= M_t\left[\{f(t) - \sum_{k=1}^{m} B_k \exp(i\lambda_k t)\}\{f^*(t) - \sum_{k=1}^{m} B_k^* \exp(-i\lambda_k t)\}\right]$$

(where the asterisk denotes the complex conjugate) then the above,

$$= M_t\{f(t)f^*(t)\} - \sum_{k=1}^{m} B_k^* M_t\{f(t) \exp(-i\lambda_k t)\}$$

$$- \sum_{k=1}^{m} B_k M_t\{f^*(t) \exp(i\lambda_k t)\} + \sum_{k_1=1}^{m} \sum_{k_2=1}^{m} B_{k_1} B_{k_1}^* M_t\{\exp(i\lambda_{k_1} t$$

$$- i\lambda_{k_2} t)\}$$

As $M_t\{\exp[i(\lambda_{k_1} - \lambda_{k_2})t]\}$ differs from zero (and is equal to 1) only for $k_1 = k_2$ the last sum reduces to the sum $\sum_{k=1}^{m} |B_k|^2$. Thus

$$M_t\{|f(t) - \sum_{k=1}^{m} B_k \exp(i\lambda_k t)|^2\}$$

$$= M_t\{|f(t)|^2\} - \sum_{k=1}^{m} B_k^* A(f,\lambda_k) - \sum_{k=1}^{m} B_k A^*(f,\lambda_k) + \sum_{k=1}^{m} B_k B_k^*$$

$$= M_t\{|f(t)|^2\} - \sum_{k=1}^{m} A(f,\lambda_k)A^*(f,\lambda_k)$$

$$+ \sum_{k=1}^{m} \{B_k - A(f,\lambda_k)\}\{B_k^* - A^*(f,\lambda_k)\}$$

$$= M_t\{|f(t)|^2\} - \sum_{k=1}^{m} |A(f,\lambda_k)|^2 + \sum_{k=1}^{m} |B_k - A(f,\lambda_k)|^2$$

and equation (2.9) is called the equation of approximation in the mean.

From the last theorem it is clear that the polynomial $\sum_{k=1}^{m} B_k \exp(i\lambda_k t)$ with fixed exponents λ_k gives the best approximation in the mean to f(t) if $B_k = A(f,\lambda_k)$ for all k, in which case we have

$$M_t\{|f(t) - \sum_{k=1}^{m} A(f,\lambda_k) \exp(i\lambda_k t)|^2\}$$

$$= M_t\{|f(t)|^2\} - \sum_{k=1}^{m} |A(f,\lambda_k)|^2$$

The left-hand side of this equation being nonnegative, it follows that

$$\sum_{k=1}^{m} |A(f,\lambda_k)|^2 \le M_t\{|f(t)|^2\} \tag{2.10}$$

This inequality is known as Bessel's inequality and is true for an arbitrary number m of real numbers λ_k. It appears that to any positive η there corresponds at most a finite number of values of λ for which $|A(f,\lambda)| > \eta$. This leads immediately to:

Theorem 2.11: *(Bohr, op.cit.)*

There are at most a countably infinite set of values of λ for which $A(f,\lambda)$ differs from zero.

Denote these values of λ by $\lambda_1, \lambda_2, \ldots,$ and write the set of these numbers as Λ.

Definition 2.3: *Set of exponents (Bohr, op.cit.)*

The set Λ for which $A(f,\lambda) \neq 0$ with $\lambda \in \Lambda$ is called the *set of exponents* of f(t) and may be written Λ_f.

Definition 2.4: *Module of f (Fink, 1974)*

The set of all real numbers which are a linear combination of the elements of Λ with integer coefficients is called the *module of f*, mod(f), i.e.

$$\text{mod}(f) = \{ \sum_{j=1}^{N} n_j \lambda_j ;\ n_j,\ N \geq 1,\ \text{integer}\}$$

That is for $f \in AP(C)$ the module of f is the smallest additive group which contains the exponents of f. The relationship between the exponents of two almost periodic functions f and $g \in AP(C)$ are contained in the following equivalent relationships (Favard, 1933; Fink, 1974):-

(i) $\text{mod}(f) \supset \text{mod}(g)$ (module containment)

(ii) for every $\eta>0$ a $\eta'>0$ exists such that $T(\eta,f(t)) \subset T(\eta',g(t))$ (translation set containment)

(iii) $T_h f = f$ implies $T_h g = g$ and that there is a $h' \subset h$ so that $T_{h'} g = g$ (assuming that $T_h f$ exists). Convergence here is either uniform, uniform on compact sets, pointwise or in the mean sense, since they are all equivalent for $g,f \in AP(C)$.

Definition 2.5: *Fourier coefficients (Bohr, op.cit.)*

The numbers $A(f,\lambda)$ for $\lambda \in \Lambda$ are called the *Fourier coefficients* of f(t).

Since Λ is countable, it can be enumerated by the positive integers and one can write the Fourier series associated with f(t) as

$$f(t) \approx \Sigma\, A(f,\Lambda)\, \exp(i\Lambda t) \tag{2.11}$$

Given the Fourier series (2.11) of $f(t) \in AP(C)$, then also

$$f^*(t) \approx \Sigma\, A^*(f,\Lambda)\exp(-i\Lambda t)$$

$$f(t+\beta) \approx \Sigma\, A(f,\Lambda)\exp(i\Lambda t)\exp(i\Lambda\beta)$$

$$\exp(i\alpha t)f(t) \approx \Sigma\, A(f,\Lambda)\exp(i(\Lambda+\alpha)t)$$

for α,β real numbers.

If f(t) is a periodic function, then the Fourier series (2.11) reduces to the usual Fourier series (2.2) for periodic functions. Of course no convergence is implied by the Fourier series representation (2.11), although since Theorem 2.10 holds for any finite set of numbers in Λ, the sum over Λ converges, that is Bessel's inequality (2.10) holds. In fact Bessel's inequality can be replaced by an equality and this gives Parseval's equation. The details of the procedure for demonstrating this are omitted (see Jensen, 1949).

Theorem 2.12: *Parseval's Equation (Jensen, 1949)*

For any $f(t) \in AP(C)$

$$M_t(|f(t)|^2) = \sum_{K=1}^{N} |A(f,\lambda_K)|^2$$

If the two almost periodic functions $f_1(t)$, $f_2(t)$ with $(f_1(t) - f_2(t)) > 0$ for all t, have the same Fourier series then from Parseval's equality applied to $f_1(t) - f_2(t) \equiv \theta(t)$, $\theta(t) \in AP(C)$, it would follow that $M_t\{|\theta(t)|^2\} = 0$. However, since $\theta(t)$ is a non-negative and non-vanishing almost periodic function it has a positive mean and therefore $M_t\{|\theta(t)|^2\} \neq 0$, and we conclude that two distinct almost periodic functions have distinct Fourier series - the so-called Uniqueness theorem.

The final aspect of the Fourier series theory of almost

periodic functions concerns their convergence, that is we ask whether or not one can actually compute the function from its corresponding Fourier series. The answer to this question has been provided by several authors, in particular Bochner (1927), by recourse to the classical approximation theorem due to Weierstrass. It is interesting to note that two problems are contained in the above discussion: one is concerned with convergence questions, while the other involves polynomial approximation. Although the two problems are closely related, they may be and have been on occasions, treated as separate. Indeed Bohr tackled the approximation problem by developing the theory of purely periodic functions of infinitely many variables - functions that can be written as:

$$f(t_1,t_2,\ldots,t_m,\ldots) \approx \sum A_k \exp[i(r_1^{(k)}\omega_1 t_1 + r_2^{(k)}\omega_2 t_2 + r_m^{(k)}\omega_m t_m + \ldots)]$$

with the $r_j^{(k)}$ rational numbers ($j = 1,2,\ldots$). A convergent sequence of such functions converges to a function which Bohr called a limit periodic function of infinitely many variables. Associated with each limit periodic function there exists a diagonal function $f(t)$ formed by setting each of the variables $t_j = t (j = 1,2,\ldots)$, that is:

$$f(t) = f(t_1,t_2,\ldots,t_m,\ldots)$$

and by means of a theorem due to Kronecker on Diophantine approximations, Bohr was able to show that the aggregate of all the values of the diagonal function $f(t)$ is everywhere dense in the aggregate of all the values of $f(t_1,t_2,\ldots,t_m,\ldots)$. Furthermore, he was able to show that $f(t)$ is almost periodic.

Unfortunately an answer to the convergence question does not come easily from Bohr's work. In fact it comes in two parts. The first stems from the classical theory of Fourier series of purely periodic functions which tells us that in the purely periodic case we should consider a more general concept than convergence, namely that of summability, because there are examples of Fourier series known to diverge at a point. Therefore, instead of considering the convergence of sequences of partial sums, for summability we consider the convergence of sequences of arithmetic means of the partial sums. Summability in this form is called *Fejer summability* and not only does it tell us something about the convergence properties of Fourier series of periodic functions but also it contains the classical theorem of Weierstrass on trigonometric polynomial approximation (see Chapter One). The second part of the answer we seek stems from the fundamental dissimilarity between purely periodic and almost periodic functions that prevents the development of the theory of the latter by *direct* analogy with the former. The problem of summation by partial sums of the Fourier series of a purely periodic function has already been examined and similar difficulties are to be expected with diagonal functions of limit periodic functions of several or an infinite number of variables. However, from the same point of view we may expect summation by arithmetic means to be applicable to the general case of almost periodic functions. Bochner *(op.cit.)* realised that the continuity of an almost periodic function implies the continuity of the corresponding limit periodic function of many variables and that this is a sufficient condition for the uniform convergence of the Fejer sums of the latter functions. He then took the diagonal functions of these Fejer sums to obtain the Fejer sums of the almost periodic function itself. The implication of Bochner's approach is the following theorem:-

Theorem 2.13: *Approximation theorem (Bochner, op.cit.)*

To any almost periodic function there corresponds a sequence of trigonometric polynomials - Bochner-Fejer polynomials - uniformly convergent to the function.

Proof: The proof of the Approximation Theorem is based on the proof for the same theorem for periodic functions. The proof for periodic functions is to show that the Cesaro-means of the partial sums converge uniformly to the continuous function. To formulate the central idea, let

$$f(t) \approx \sum_{|\nu|<\infty} A_\nu \exp[i\nu\omega t]$$

then the Cesaro means (Fejer sums) are given by the formula

$$\sigma_k(t) = \sum_{|\nu|<k} \left[1 - \frac{|\nu|}{k}\right] A_\nu \exp[i\nu\omega t]$$

Since $A_\nu = M_s\{f(s)\exp[-i\nu\omega s]\}$, it follows that

$$A_\nu \exp[i\nu\omega t] = M_s\{f(s)\exp[-i\nu\omega(s-t)]\}$$

$$= M_s\{f(s+t)\exp[-i\nu\omega s]\}$$

whence

$$\sigma_k(t) = M_s\{f(s+t)\Pi_k(\omega s)\}$$

where $\Pi_k(s)$ is the Fejer kernel and is defined by the equation

$$\Pi_k(s) = \sum_{|\nu|<k}\left(1 - \frac{|\nu|}{k}\right)\exp[-i\nu s] = \frac{1}{k}\left(\frac{\sin\frac{k}{2}s}{\sin\frac{s}{2}}\right)^2$$

The main properties of $\Pi_k(s)$ are

(i) $\Pi_k(s) \geq 0$

and (ii) $M(\Pi_k) = 1$.

Property (i) is clear from the second representation and (ii) from the first since $M(\Pi_k)$ is the constant term. The proof then is as follows,

$$|f(t) - \sigma_k(t)| = M_s\{f(t)\Pi_k(\omega s) - f(s+t)\Pi_k(\omega s)\}$$

$$\leq M_s\{|f(t) - f(s+t)|\Pi_k(\omega s)\}$$

and the idea is that for $|s|$ small the first factor is small and for $|s|$ bounded away from 0, $\Pi_k(\omega s)\to 0$ as $k\to\infty$. To solve the almost periodic case, products of Π_k's are taken, one for every periodic component of the almost periodic function. That is, a finite product of the kernels is

$$\Pi_{\substack{k_1,k_2,\ldots,k_m \\ \omega_1,\omega_2,\ldots,\omega_m}}(s) = \Pi_{k_1}(\omega_1 s)\Pi_{k_2}(\omega_2 s)\ldots\Pi_{k_m}(\omega_m s)$$

$$= \sum_{|\nu|<k}\left[1-\frac{|\nu_1|}{k_1}\right]\left[1-\frac{|\nu_2|}{k_2}\right]\ldots\left[1-\frac{|\nu_m|}{k_m}\right]$$

$$\times \exp[-i(\nu_1\omega_1 + \nu_2\omega_2 + \ldots + \nu_m\omega_m)s]$$

where $|\underline{\nu}| < |\underline{k}|$ denotes all the summations over $|\nu_1|<k_1,\ldots,|\nu_m| < k_m$, and is taken, for $\omega_1,\omega_2,\ldots,\omega_m$, certain real linearly independent numbers. This composite kernel has the same characteristic properties as the Fejer kernel - it is never negative and its mean value is equal to 1. This kernel is called the Bochner-Fejer kernel and using it we can form a Bochner-Fejer polynomial by direct analogy with the purely periodic case, that is,

$$BF_{\substack{k_1,k_2,\ldots,k_m \\ \omega_1,\omega_2,\ldots,\omega_m}}(t) = M_s\{f(s+t)\Pi_{\substack{k_1,k_2,\ldots,k_m \\ \omega_1,\omega_2,\ldots,\omega_m}}(s)\}$$

where f(t) is now an almost periodic function,

$$f(t) \approx \sum A_\nu \exp[i\lambda_\nu t]$$

Thus we can write $BF_{\substack{k_1,k_2,\ldots,k_m \\ \omega_1,\omega_2,\ldots,\omega_m}}(t)$ as

$$= \sum_{|\underline{\nu}|<\underline{k}}\left[1-\frac{|\nu_1|}{k_1}\right]\left[1-\frac{|\nu_2|}{k_2}\right]\ldots\left[1-\frac{|\nu_m|}{k_m}\right] \times$$

$$A(f,\nu_1\omega_1+\nu_2\omega_2+\ldots+\nu_m\omega_m)\exp[i(\nu_1\omega_1+\nu_2\omega_2+\ldots+\nu_m\omega_m)t], \quad (2.12)$$

where as usual

$$A(f,\Lambda) = M_t\{f(t)\exp[-i\Lambda t]\},$$

and each $\lambda_\nu \in \Lambda$ is written as

$$\lambda_\nu = \nu_1\omega_1 + \nu_2\omega_2 + \ldots + \nu_m\omega_m$$

In this representation for λ_ν the numbers $\nu_1,\nu_2,\ldots,\nu_m$ are rationals and the linearly independent set of numbers $\omega_1,\omega_2,\ldots,\omega_m$ are called a *base* for Λ_f. It follows from (2.12) that the exponents of $BF_{\substack{k_1,k_2,\ldots,k_m \\ \omega_1,\omega_2,\ldots,\omega_m}}(t)$ are contained in the set of exponents of $f(t)$.

Introducing a simpler notation, the Bochner-Fejer polynomials are written in the form

$$BF_k(t) = \Sigma\, d_\nu^{(k)}\, A_\nu \exp[i\lambda_\nu t],$$

where the $d_\nu^{(k)}$ satisfy the inequality $0 \leq d_\nu^{(k)} \leq 1$ and only a finite number of them are different from zero. We observe that $d_\nu^{(k)}$ depend only on $k_1,k_2,\ldots,k_m$, $\omega_1,\omega_2,\ldots,\omega_m$ and on λ_ν but not on the values of the coefficients A_ν. Bochner has shown (Bochner, 1927) that the set of all Bochner-Fejer polynomials together with the set of $f(t)$ are uniformly continuous and uniformly almost periodic. Under these conditions, our main problem of finding a sequence of Bochner-Fejer polynomials uniformly convergent to the function $f(t)$ is equivalent to finding a sequence convergent in mean. Thus the problem is given an $\xi > 0$ to find a Bochner-Fejer polynomials $BF_k(t) = BF_{\substack{k_1,k_2,\ldots,k_m \\ \omega_1,\omega_2,\ldots,\omega_m}}(t)$ such that

$$M_t\{|f(t) - BF_k(t)|^2\} < \xi.$$

Recall that the Fourier series of $f(t)$ is

$$F(t) \approx \sum_{\nu=1}^{\infty} A_\nu \exp[i\lambda_\nu t]$$

and let $\alpha_1, \alpha_2, \ldots$ be a base of Λ_f. Define ν_o so that

$$\sum_{\nu=\nu_o+1}^{\infty} |A_\nu|^2 < \frac{\xi}{2}. \tag{2.13}$$

Let m be the largest index of α's in the linear expressions of $\lambda_1, \lambda_2, \ldots, \lambda_{\nu_o}$, so that we can write

$$\lambda_\nu = r_1^{(\nu)} \alpha_1 + r_2^{(\nu)} \alpha_2 + \ldots + r_m^{(\nu)} \alpha_m, \qquad (\nu = 1, 2, \ldots, \nu_o)$$

where all r's are rational. Let q be the common denominator of all the numbers $r_j^{(\nu)}$ $(j = 1, 2, \ldots, m;\ \nu = 1, 2, \ldots, \nu_o)$. We write

$$\lambda_\nu = R_1^{(\nu)} \frac{\alpha_1}{q} + R_2^{(\nu)} \frac{\alpha_2}{q} + \ldots + R_m^{(\nu)} \frac{\alpha_m}{q}$$

where all $R_j^{(\nu)}$ are integers. Let R be the maximum of all $|R_j^{(\nu)}|$. Define numbers $\phi > 0$ and $N > 0$ by the conditions

$$\phi^2 \sum_{\nu=1}^{\nu_o} |A_\nu|^2 < \frac{\xi}{2} \tag{2.14}$$

$$\left(1 - \frac{R}{N}\right)^m > 1 - \phi \tag{2.15}$$

Take $\omega_j = \frac{\alpha_j}{q}$ $(j=1,2,\ldots,m)$ and all $k_1, k_2, \ldots, k_m$ greater than N. By (2.12), $BF_k(t) = BF_{\substack{k_1, k_2, \ldots, k_m \\ \omega_1, \omega_2, \ldots, \omega_m}}(t)$

$$= \sum_{\nu=1}^{\nu_o} \left[1 - \frac{|R_1^{(\nu)}|}{k_1}\right] \cdots \left[1 - \frac{|R_m^{(\nu)}|}{k_m}\right] A_\nu \exp|i\lambda_\nu t|$$

$$+ \sum_{\nu=\nu_o+1}^{\infty} d_\nu A_\nu \exp[i\lambda_\nu t]$$

where d_ν differs from zero only for a finite number of values of ν and $0 \le d_\nu \le 1$ for all $\nu > \nu_o$.

We have

$$M_t\{|f(t) - BF_k(t)|^2\}$$

$$= \sum_{\nu=1}^{\nu_o} \left\{1 - \left[\left[1 - \frac{|R_1^{(\nu)}|}{k_1}\right]\cdots\left[1 - \frac{|R_m^{(\nu)}|}{k_m}\right]\right]\right\}^2 |A_\nu|^2$$

$$+ \sum_{\nu=\nu_o+1}^{\infty} (1-d_\nu)^2 |A_\nu|^2$$

By (2.13), (2.14) and (2.15)

$$M_t\{|f(t)-BF_k(t)|^2\} \le \phi^2 \sum_{\nu=1}^{\nu_o} |A_\nu|^2 + \sum_{\nu=\nu_o+1}^{\infty} |A_\nu|^2 < \xi$$

Therefore $BF_k(t)$ is a polynomial of the kind required and this concludes the proof of Theorem 2.13.

Note that Bochner's procedure not only proves the existence of trigonometric polynomials $BF_k(t)$ such that $\sup_t|f(t)-BF_k(t)| < \eta$ for some $\eta > 0$ as small as we please, but also it gives a definite algorithm for finding the $BF_k(t)$. Furthermore, the polynomials $BF_k(t)$ have exponents which are Fourier exponents of $f(t)$.

An obvious corollary to theorem 2.13 is that almost periodic functions are precisely those functions that can be uniformly approximated by trigonometrical polynomials. Since we are in the main concerned with almost periodic differential equations, questions of differentiability and integration of Fourier series of almost periodic functions are of special interest. Suppose that $f(t)$ and $f'(t) \in AP(C)$, the Fourier series of $f'(t) \in AP(C)$ is just the formal derivative of the Fourier series of $f(t)$, i.e.

$$A(f',\lambda_K) = i\,\lambda\,A(f,\lambda_K), \qquad i = \sqrt{-1}.$$

Given $f(t) \in AP(C)$, the simplest condition (Meisters, 1958) on the Fourier series of f(t) which yields $\int_0^t f(s)ds \in AP(C)$ is

$$\sum_k |A(f,\lambda_k)\lambda_k^{-1}| < \infty$$

since

$$F(t) = \int_0^t f(s)ds = A_0 + \sum_k A(f,\lambda_k)(i\lambda_k)^{-1} \exp(i\lambda_k t)$$

Although the numbers λ_k occur in the denominator of the Fourier series of F(t), it does not effect the validity of the series since $\lambda \neq 0$. That is for $F(t) \in AP(C)$ it is necessary (but not sufficient) that the Fourier exponents λ_k of f(t) are non-zero.

In anticipation of the main theme in later chapters concerning differential equations whose coefficients are almost periodic functions containing a parameter, in the next section we examine the fundamental properties of this class of functions.

2.4 Almost Periodic Functions Depending Uniformly on a Parameter

In the study of vector differential equations with almost periodic coefficients, the properties of vector continuous complex valued functions dependent upon a parameter vector x is most important. Fortunately almost all of the properties of almost periodic functions discussed in sections 2.2 and 2.3 can easily be extended to almost periodic functions dependent upon a parameter vector. Consider the n-vector continuous functions $f(t,x)$ $(f: R \times D \to E^n)$ where D is an open subset in E^n (more generally a separable Banach space) and $x \in D$.

Definition 2.6: *Almost periodic functions dependent upon a parameter*

A function f(t,x) is called *almost periodic* in t uniformly with respect to $x \in D$ if for any $\eta > 0$ and compact set $F \subset D$ there exists a positive number $\ell_\eta(F)$ such that any interval of the real line of length $\ell_\eta(F)$ contains a τ for which $|f(t+\tau,x) - f(t,x)| < \eta$ for all $t \in R$ and for all $x \in F$. In this case we denote $f \in AP(E^n)$.

Similarly following definition 2.1, the number η is called a η-translation number of $f(t,x)$ and we denote by $T(\eta,f,F)$ the set of all η-translation of functions dependent upon a parameter which are identical to those discussed in section 2.2. Note that the translation numbers (almost periods) are again selected from a relatively dense set.

Almost periodic functions dependent upon a parameter have a variety of continuity and algebraic properties which are readily derived from the above definition. For example if $f \in AP(E^n)$ then $f(t,x)$ is bounded and uniformly continuous on $R \times F$, F any compact subset in D ($x \in D$). These boundedness and continuity conditions can be established by setting $\eta = 1$ and selecting an interval $\ell_\eta = \ell(\frac{\eta}{3},f)$ for $\eta > 0$. Also for vector valued functions $f(t,x) = \{f_i(t,x)\} \in AP(E^n)$ each component $f_i(t,x) \in AP(E)$, and conversely. Similar algebraic results to those of theorems 2.3-2.6 also hold for the components f_i of $f(t,x) \in AP(E^n)$: i.e. if each $f_i(t,x) \in AP(E)$ and for any $g(t,x) \in AP(E^n)$ then $\alpha f_i(t,x)$, $f_i^2(t,x)$, $f_i(t,x) + g(t,x)$ and $f_i(t,x) \times g(t,x)$ are all almost periodic in t uniformly for $x \in D$, for some constants α and for all i. Moreover if

$$\inf_{\substack{t\in R \\ x\in F}} |g(t,x)| > 0, \qquad F \subset D,$$

then $f_i(t,x)g(t,x)^{-1}$ is almost periodic in t uniformly for $x \in D$. Some of the above results are given more formly in Theorems 2.14-2.16 since they are of vital importance in the development of polynomial approximations of almost periodic functions dependent upon a parameter. The question of integrability of functions $f(t,x) \in AP(E^n)$ will be dealt with in Chapters Four and Six.

The next two theorems are of vital importance to the development presented in Chapter Six. They define properties of almost periodic functions containing a parameter which are essential in the problem of polynomial approximation. By analogy with Theorems 2.1 and 2.2 we have:

Theorem 2.14: *(Corduneanu, 1968)*

If D is a compact set in E^n, then the function $f(t,x) \varepsilon AP(E^n)$ is almost periodic in t uniformly with respect to x, is bounded on R×D.

Proof: uses the same argument as in Theorem 2.1. The details are omitted.

Theorem 2.15: *(Corduneanu, 1968)*

Under the same hypothesis as in Theorem 2.18, it follows that $f(t,x) \varepsilon AP(E^n)$ is uniformly continuous on the set R×D.

Proof: Essentially the same as in Theorem 2.2. The details are omitted.

The next theorem we present in this section is analogous to Theorem 2.4 for almost periodic functions without parameters, and tells something of the properties of convergent sequences of almost periodic functions containing a parameter.

Theorem 2.16: *(Corduneanu, 1968)*

If a sequence of almost periodic functions $\{f_k(t,x)\}$ uniformly dependent on the parameter $x \varepsilon D$ is uniformly convergent on R×D to the function f(t,x), then f(t,x) is also almost periodic in t uniformly with respect to $x \varepsilon D$.

Proof: Given η, there exists a function $f_{k_0}(t,x)$ such that

$$|f(t,x) - f_{k_0}(t,x)| < \frac{\eta}{3}, \quad \text{for all} \quad (x,t) \varepsilon D\times R.$$

Let now x be an almost period from the set of translation numbers $T(\frac{1}{3}\eta, f_{k_0}(t,x))$. Then

$$|f(t+\tau,x) - f(t,x)| \leq |f(t+\tau,x) - f_{k_0}(t+\tau,x)|$$
$$+ |f_{k_0}(t+\tau,x) - f_{k_0}(t,x)| + |f_{k_0}(t,x) - f(t,x)| < \eta$$

which shows that

$$T(\eta, f(t,x) \supset T(\tfrac{1}{3}\eta, f_{k_o}(t,x))$$

and this proves the theorem.

The relationship between the exponents of two almost periodic functions $f(t,x)$, $g(t,x)$ dependent upon a parameter vector x, is given in the following theorem due to Favard (1933):

Theorem 2.17: *Module containment*

Let $f(t,x)$ and $g(t,x)$ be almost periodic in t uniformly for $x \in D$, then for any compact set $F \subset D$, $\text{mod}(f,F) \supset \text{mod}(g,F)$ implies $\text{mod}(f,D) \supset \text{mod}(g,D)$.

Proof: Let $T_h f(t,x) = f(x,t)$ uniformly on $R \times D$ then by the equivalent statements on module containment following definition 2.6. $T_h g = g$ uniformly, and also $T_h f(g,t) = f(g,t)$ so that $\text{mod}\, f(g,t) \subset \text{mod}(f)$ or equivalently $\text{mod}(f,D) \subset \text{mod}(g,D)$.

The fourier series theory for functions $f(t,x) \in AP(E^n)$ dependent upon a parameter $x \in D \subset E^n$, is in general a direct extension of the Bohr transform theory of section 2.3. The parallel Bohr transform of an almost periodic function to that of (2.3) for $f(t,x) \in AP(E^n)$ uniformly for $x \in D$, a compact subset of E^n is

$$A(f,\lambda,x) = \lim_{W\to\infty} \frac{1}{W} \int_0^W f(t,x) \exp(-i\lambda t)dt \tag{2.16}$$

where for each (λ,f,x), $A(f,\lambda,x) \in E^n$, and for any $\lambda \in R$, $A(f,\lambda,x)$ exists uniformly for x in compact sets and is continuous in x. This latter condition on the fourier coefficient $A(f,\lambda,x)$ follows directly from Theorem 2.11. Hence we have a formal correspondence between an almost periodic function $f(t,x)$ and its fourier series

$$f(t,x) \approx \sum_{K=1}^{\infty} A(f,\lambda_K,x) \exp(i\lambda_K t) \tag{2.17}$$

The definitions for sets of exponents and the module of $f(t,x) \in AP(E^n)$ follow identically to those of definitions 2.3 and 2.4.

Finally we note that both Bessel's inequality and Parseval's equation hold in the mean for any $f(t,x) \varepsilon AP(E^n)$.

2.5 Bochner's Criterion

In certain applications, notably in the study of differential equations, Bochner (1927) found it useful to characterise almost periodic functions by means of a compactness criterion. This definition plays an essential role in the general theory of almost periodic functions and leads to results which are not easily derived from the Bohr definition. The starting point consists of considering together with a given continuous function $f:R \to C$ the set of its *translates* $\{f_\tau : f_\tau(t) = f(t+\tau),\ \ \tau \varepsilon R\} = S(f)$ and its closure $\overline{S(f)}$ in the topology of uniform convergence.

Definition 2.7: *Almost periodic function (Bochner, 1927)*

The continuous function $f:R \to C$ is almost periodic if from every sequence of real numbers $\{h_n\}$ one can extract a subsequence $\{h_{n_k}\}$ such that $\lim_{k\to\infty} f(t+h_{n_k}) = g(t)$ exists uniformly on R.

To facilitate further discussion and remove the necessity of writing double subscripts, we introduce the following notation for sequences: if $\alpha = \{\alpha_n\}$ and $\beta = \{\beta_n\}$, then $\alpha + \beta = \{\alpha_n + \beta_n\}$. Also $\beta \subset \alpha$ will mean that $\{\beta_n\}$ is a subsequence of $\{\alpha_n\}$. If $\alpha \subset \alpha'$ and $\beta \subset \beta'$, we say that α and β are *common subsequences* of α' and β' if $\{\alpha_k\} = \{\alpha_{n_k}\}$ and $\{\beta_k\} = \{\beta_{n_k}\}$ for a common set of indices $\{n_k\}$. Finally we define the *translation operator* $T_h f = \lim_{k\to\infty} f(t+h_k)$ which notationally simplifies definition 2.7 to:

Definition 2.8: *Almost periodic function (Fink, 1974)*

The continuous function $f:R \to C$ is almost periodic if for every sequence h' there exists an $h \subset h'$ such that $T_h f$ exists uniformly.

Not only is this definition equivalent to Bohr's definition of almost periodic functions, it is also the definition of f(t) being *normal* (Bochner, 1927) and each of these definitions implies the characteristic properties of the others. The equivalence of the Bochner definition of almost periodicity (given essentially by the space AP(C)) and the Bohr definition is summarised in the following theorem:-

Theorem 2.18: *(Fink, 1974)*

For any continuous complex valued function $f:R\to C$, then $f \in AP(C)$ if and only if $f(t)$ is Bohr almost periodic.

It is clear that through the use of the translation operator we are able to generate new almost periodic functions from $f(t)$. The collection of *all* such functions g such that there is an h for which $T_h f = g$ uniformly is called the *hull* of f and is denoted by H(f). For general almost periodic functions such as $f(t) = \sin t + \sin \sqrt{2}\, t$ in H(f), g can be shown not to be a translate of f. It is not difficult to show that H(f) is a metric space which is compact in the uniform norm if and only if $f(t) \in AP(C)$, this is equivalent to the set of translates $S(f) = \{f_\tau : f_\tau(t) = f(t+\tau),\ \tau \in R\}$ being totally bounded (Bochner and Von Neumann, 1935). Clearly from the definition of the hull, H(f), if $f \in AP(C)$ then for any $g \in H(f)$, $H(g) = H(f)$.

The space AP(C) is an algebra over C, since for f and g both $\in AP(C)$, $T_h(f+g) + T_h g = T_{h'} f + T_{h'} g$ uniformly for $h \subset h'$ and thence $(f+g) \in AP(C)$, also $T_h(fg) = T_h(f)T_h(g)$ exists uniformly so that AP(C) is closed under products. Then clearly finite sums of periodic functions are almost periodic functions and we may conclude that AP(C) is a complete metric space (Banach space) that contains all periodic functions. It is in fact the smallest completely normed space with this property and most importantly for our purposes, AP(C) is closed under differentiation and integration (Corduneanu, 1968). The differentiation property follows simply from the fact that all $f \in AP(C)$ are bounded and uniformly continuous, then if the derivative f' exists everywhere then $f' \in AP(C)$.

To complete this section we present a further definition of almost periodicity which relies only on *pointwise convergence*, although uniform convergence is implied.

Definition 2.9: *Almost periodic function (Bochner, 1962)*

The continuous function $f:R \to C$ is almost periodic if from every pair of sequences h_1', h_2' one can extract common subsequences $h_1 \subset h_1'$ and $h_2 \subset h_2'$ such that

$$T_{h_1} T_{h_2} f = T_{h_1+h_2} f$$

pointwise.

The above condition is both a necessary and sufficient condition for convergence whether it be pointwise or uniform on compact sets. The utility of Bochner's pointwise definition for almost periodicity is revealed in Chapter Six in the context of differential equations with almost periodic coefficients.

2.6 Limiting Cases of Almost Periodic Functions

The concept of asymptotically almost periodic functions f(t) defined on the half real line $R_+ = [0,\infty)$ with values in E^n $(f:R_+ \to C^n)$ was first introduced by Frechet (1941) and dealt with in some depth by Corduneanu (1968). For completeness we introduce its main properties.

Definition 2.10: *Asymptotically almost periodic functions*

A continuous valued function $f:R_+ \to C^n$ is said to be *asymptotically almost periodic* if it is the sum of a continuous almost periodic function $p(t) \in E^n$ and a continuous function $q:R_+ \to C^n$ which tends to zero as $t \to \infty$. i.e.

$$f(t) = p(t) + q(t) \in AAP(C^n)$$

if $p(t) \in AP(C^n)$, $q:R_+ \to C^n$ and $\lim_{t\to\infty} q(t) = 0$.

This decomposition is unique if $f(t) \in AAP(C^n)$ and f(t) is bounded and uniformly continuous on $R_+ = [0,\infty)$. Equally, if f and its derivative f' are both $AAP(C^n)$ then the decomposition of the derivative f',

$$f'(t) = p'(t) + q'(t)$$

is also unique with $p'(t) \in AP(C^n)$.

We can relate the above definition of asymptotically almost periodicity to the Bohr and Bochner definitions of almost periodicity through the following definitions:

Definition 2.11: *Property P (Corduneanu, 1968)*

$f:R_+ \to C^n$ has *property P* if given an $\eta > 0$ there is a $\ell_\eta > 0$ and a $T(\eta) \geq 0$ such that every interval of length ℓ_η on R_+ contains a τ such that

$$|f(t+\tau) - f(\tau)| < \eta \qquad \text{for} \quad t \geq T(\eta).$$

Definition 2.12: *Property L (Corduneanu, 1968)*

$f:R_+ \to C^n$ has *property L* if for any real sequence $\{h'_k\}$ such that $h_k' > 0$ and $h_k' \to \alpha$ as $k \to \infty$, we can select a subsequence $\{h_k\} \subset \{h_k'\}$ such that $f(t+h_k)$ converges uniformly on R_+, i.e.

$$\lim_{k\to\infty} f(t+h_k) = g(t) \qquad \text{for} \quad t \in R_+.$$

Definitions 2.10-2.12 are equivalent (Yoshizawa, 1975) so that any $f:R_+ \to C^n$ which has either property P or L is also asymptotically almost periodic.

A special case of almost periodic functions dependent upon a parameter are quasi-periodic functions $f(t,x)$ $(f:R\times E^n \to E^n)$ (Nakajima, 1972). Let I_j be a unit vector in D^k such that the jth component is 1 and the others are zero. Let I be a vector in E^k such that all of the components are one $(k \leq n)$.

Definition 2.13: *Quasi-periodic functions*

The function $f(t,x)$ $(f:R\times E^n \to E^n)$ is said to be *quasi-periodic* in t if there is a finite number of non-zero real numbers w ,w $\ldots w_k$, and a function $F(z,x)$ where $F(z,x)$ $(F:R\times E^n \to E^n)$ is such that

$$F(z+w_jI_j,x) = F(z,x) \qquad \text{for all} \quad z \in E^k,$$

$x \in D$, and $j = 1,2,\ldots n$

and

$F(tI,x) = f(t,x)$ for all $t \varepsilon R$ and $x \varepsilon D$ where D is an open subset in E^n.

The most important result on quasi-periodic functions for our purposes is the existence of a fourier series:-

Theorem 2.19: *(Nakajima, 1972)*

Let $f(t,x)$ $(f:R\times E^n \to E^n)$ for $x \varepsilon D$ an open subset in E^n. The function $f(t,x)$ is quasi-periodic in t if and only if it is almost periodic in t uniformly for $x \varepsilon D$ and its module, $\mathrm{mod}(f) = \left\{\frac{2\pi}{w_1},\ldots,\frac{2\pi}{w_k}\right\}$ has a finite integrable base. That is $f(t,x) \varepsilon QP(E^n)$ is almost periodic with fourier series

$$f(t,x) \approx \sum_m A_m(x,\lambda) \exp\left\{2\pi it\left\{\frac{m_1}{w_1} + \ldots + \frac{m_k}{w_k}\right\}\right\}, \quad i = \sqrt{-1}$$

where $w_1,\ldots,w_k$, are real positive numbers, $m = (m_1,\ldots,m_k)$ for integer m_i, and

$$A_m(x,\lambda) \triangleq \lim_{w\to\infty} \frac{1}{w}\int_0^w f(t,x)\exp(-\lambda t)dt \neq 0.$$

2.7 Notes

The functions studied by Bohl, the so-called quasi-periodic functions, are a special class of almost periodic functions whose structure closely resembles the limit periodic functions introduced by Bohr. Perhaps this observation conditioned Bohr's original thinking and led to his investigation of diagonal functions of limit periodic functions of many variables. Work similar to Bohl's was performed by M. E. Esclangon and is reported in two readily accessible papers: "Sur une extension de la notion de periodicité", Comptes Rendus Acad. Sci., Paris, 1902, and "Sur les fonctions quasi-périodiques moyennes déduites d'une fonction quasi-périodique", Comptes Rendus Acad. Sci., Paris, 1913.

The early papers by Bohr, "Sur les fonctions presque périodiques" and "Sur l'approximation des fonctions presque périodiques par des sommes trigonométriques", both in Comptes Rendus Acad. Sci., Paris, 1925, seem to substantiate the view that he was inspired by Bohl's and Esclangon's work; the following extract is taken from the second paper

> ... les recherches très intéressantes de ces géomètres (Bohl and Esclangon - our brackets), les fonctions quasi périodiques peuvent se representer par la forme $f(t) = f(t,t,\ldots,t)$, ou $f(t_1,t_2,\ldots,t_m)$ est une fonction continue des variables $t_1,t_2,\ldots,t_m$, qui est rigoureusment périodique par rapport à chacune de ses m variables. C'est à l'aide de cette représentation que Bohl a obtenu son résultat principal, que les fonctions quasi périodiques sont identiques aux fonctions f(t), qu'on peut approcher d'une manière uniforme par des sommes trigonométriques de la forme $\Sigma B_k \exp|i(k_1\omega_1 + k_2\omega_2 + \ldots + k_m\omega_m)t|$.
>
> Pour les fonctions presque périodiques dont les exposants λ_k possèdent une base entière quelconque, on peut généraliser le résultat de Bohl et Esclangon de la manière suivante:
>
> Tout fonction presque périodique f(t) à laquelle appartiennent des exposants λ_k avec une base entière, peut se representer sous la forme
>
> $f(t) = f(t,t,\ldots,t,\ldots)$
>
> où $f(t_1,t_2,\ldots,t_m,\ldots)$ est une fonction uniformement continue d'une infinité de variables réeles, qui est rigoureusement périodique par rapport à toutes les variables.

The Fourier series theory of almost periodic functions was developed soon after Bohr's original work, and apart from Bohr's own efforts in this direction, a valuable paper is S. Bochner, "Properties of Fourier series of almost periodic functions", Proc.

Lond. Math. Soc. (2), **26**, 433 (1927) in which the tests for convergence and summability, arithmetic properties and polynomial approximation has been dealt with by many authors. In his paper "Localisation of best approximation", Ann. Math. Stud. 25, Princeton, 1950, S. Bochner investigates polynomial approximation by considering the convergence and summability of Fourier series.

Throughout this chapter little has been said about the almost periodicity of integrals of almost periodic functions since, in general terms, little is known about the problem. Certainly in Euclidean spaces the boundedness of the integral is necessary and sufficient to ensure its almost periodicity, although extension to general metric spaces is difficult. For more on this see the book by L. Amerio and G. Prouse, "Almost periodic functions and functional equations", Van Nostrand Reinhold, New York, 1971. For our purposes the most easily interpreted conditions for the almost periodicity of such an integral are derived from the Fourier series theory. The conditions of such an integral are derived from the Fourier series theory. The conditions are not simple and there are many of them. A recent comment on this is by G. H. Meisters, "On the almost periodicity of the integral of an almost periodic function", Am. Math. Soc. Notices, **5**, 683 (1958), which may be consulted with profit. An earlier paper which is also useful is by S. Bochner, "Remark on the integration of almost periodic functions", J. Lond. Math. Soc., **8**, 250 (1933), which is really a comment on the differential equation $\dot{x}(t) = f(t)$ with $f(t)$ almost periodic.

Generalisations of Bohr's theory have appeared and arise from different forms of convergence. A good review of this development is A. M. Fink, "Almost periodic functions invented for specific purposes", SIAM Review, **14**, 572 (1972). To conclude this section it is certainly worth noting that one of the richest sources of material on almost periodic functions is "Collected Mathematical Works" of H. Bohr, Copenhagen, 1952.

References

Amerio, L. and Prouse, G. (1971). "Almost Periodic Functions and Functional Differential Equations", Van Nostrand and Reinhold, New York

Besicovich, A.S. (1932). "Almost Periodic Functions", Cambridge Univ. Press

Bochner, S. (1927). *Math.Ann.*, **96**, 119

Bochner, S. (1927). *Proc.London Math.Soc.Ser.2*, **26**, 43

Bochner, S. (1933). *J.London Math.Soc.*, **8**, 250

Bochner, S. (1950). *Ann.Math.Stud.*, **25**, Princeton Univ. Press

Bochner, S. (1962). *Proc.Nat.Acad.Sci.*, **48**, 2039-2043

Bochner, S. and Von Neumann, J. (1935). *I.Ann.Math.*, **36**, 255-290

Bohl, P. (1906). *J.fur.Reine U.Angew.Math.*, **131**, 268

Bohr, H. (1924). *Acta Math.*, **46**, 101

Bohr, H. (1925). "Sur les Fonctions Presque Périodiques", Comptes Rendus Acad.Sci., Paris

Bohr, H. (1925). "Sur l'approximation des Fonctions Presque Périodiques par des Sommes Trigonométriques", Comptes Rendus Acad.Sci., Paris

Bohr, H. (1952). "Collected Mathematical Works", Copenhagen

Corduneanu, C. (1968). "Almost Periodic Functions", Wiley Interscience, New York

Esclangon, M.E. (1902). "Sur les extension de la Notion de Periodicité", Comptes Rendus Acad.Sci., Paris

Esclangon, M.E. (1913). "Sur les Fonctions Quasi-périodiques Moyennes Déduites d'une Fonctions Quasi-periodiques", Comptes Rendus Acad.Sci., Paris

Favard, J. (1933). "Leçons sur les Fonctions Presque Périodiques", Gauthier Villars, Paris

Fink, A.M. (1972). *SIAM Review*, **14**, 572

Fink, A.M. (1974). "Almost Periodic Differential Equations", Springer Verlag lecture notes in Mathematics, No.377, New York

Frechet, M. (1941). *Rev.Scientifique*, **79**, 341-354

Jensen, B. (1949). *Det.KGL, Dan Ke Viden. Selskob.Mat-Fys.Meddel*, **25**, 1-12

Meisters, G.H. (1958). *Amer.Math.Soc.*, **5**, 683

Nakajima, F. (1972). *Funkcial Ekvac.*, **15**, 61-73

Yoshizawa, T. (1975). "Stability Theory and the Existence of Periodic Solutions and Almost Periodic Solutions", Appl.Maths. Sci., No.14, Springer Verlag, New York

Chapter 3

PROPERTIES OF ORDINARY DIFFERENTIAL EQUATIONS

3.1 Introduction

Continuous dynamical systems with a finite dimensional state vector are often represented by an ordinary differential equation. A real scalar t (called time) and an open set D in R^{n+1} are involved, where R^n is the space of real n-dimensional column vectors. An element of D is written (x,t). Let $f:D \to R^n$ be continuous and let $\dot{x}$ denote dx/dt. Then a general differential equation of the type described is a relation of the form

$$\dot{x}(t) = f(t,x(t)), \tag{3.1}$$

or briefly

$$\dot{x} \doteq f(t,x)$$

If x is a continuously differentiable function defined on $J \subset R$, $(x,t) \in D$, $t \in J$ and x satisfies (3.1), we say that x is a *solution* of (3.1) on an interval J.

Given $(x_o,t_o) \in D$, the problem of finding an interval J containing t_o and a solution x of (3.1) satisfying $x(t_o) = x_o$ is called an *initial value problem* for (3.1). If there exists a solution to the initial value problem, we refer to this as the solution of (3.1) passing through (x_o,t_o). Symbolically the initial value problem is written

$$\dot{x} = f(t,x), \qquad x(t_o) = x_o, \qquad t \in J \tag{3.2}$$

which is equivalent to

$$x(t) = x_o + \int_{t_o}^{t} f(s,x(s))ds \tag{3.3}$$

provided that $f(t,x)$ is continuous

3.2 Existence and Uniqueness of Solution

Our main concern in this section is to establish general theorems about the existence and uniqueness of solutions of the initial value problem (3.2). The general questions of continuation of solutions and continuous dependence of solutions on initial data are not considered here; the interested reader should consult Coddington & Levinson (1955) for the relevant discussion and technical details.

The presentation in this section is adapted from Coddington & Levinson *(op.cit.)* Hale (1969), Hille (1969), Rosenbrock & Storey (1970), Curtain & Pritchard (1977) and Birkhoff & Rota (1978). The first theorem we shall derive asserts the existence of at least one solution to the initial value problem (3.2) if $f(t,x)$ is continuous in D.

Theorem 3.1: *Existence (Peano)*

Suppose that f is defined and continuous in $D = \{(x,t): \|x-x_o\| < \beta, \ x \in R^n; \ |t-t_o| < \alpha, \ t \in R\}$, also suppose that $\|f(x,t)\| \leq m$ in D. Then there is at least one solution of (3.1) passing through (x_o,t_o) which is defined on the interval $[t_o-r, t_o+r]$ where

$$r < \min(\alpha, \beta m^{-1}) \tag{3.4}$$

Proof: Hale *(op.cit.)* gives a proof which uses the Schauder fixed point theorem (see Chapter One). The following proof is taken from Hille *(op.cit.)*, who demonstrates existence on the interval $[t_o,t_o+r]$. The extension to $[t_o-r,t_o]$ is immediate. Divide the interval $[t_o, t_o+r]$ into 2^k equal parts and set

$$t_{jk} = t_o + j2^{-k}r \tag{3.5}$$

Define a sequence of piecewise linear functions $\{x_k(t)\}$ by the following formulae:

$$x_k(t) = x_o + f(t_o, x_o)(t-t_o) \qquad t_o \le t \le t_{1k},$$

$$x_k(t) = x_k(t_{jk}) + f(t_{jk}, x_k(t_{jk}))(t-t_{jk}) \qquad t_{jk} \le t \le t_{j+1,k} \tag{3.6}$$

where $j = 1,2,3,\ldots,2^k-1$.

We now show that the functions $x_k(t)$ are uniformly bounded and equicontinuous in $[t_o, t_o+r]$. In the jth interval

$$\|x_k(t)-x_k(t_{jk})\| \le \|f(t_{jk}, x_k(t_{jk}))\|2^{-k}r \le m2^{-k}r \tag{3.7}$$

which gives

$$\begin{aligned}\|x_k(t)-x_o\| &\le \|x_k(t)-x_k(t_{jk})\| + \|x_k(t_{jk})-x_k(t_{j-1,k})\| \\ &\quad + \|x_k(t_{j-1,k})-x_k(t_{j-2,k})\| + \ldots + \|x_k(t_{1k})-x_o\| \\ &\le (j+1)m2^{-k}r \le mr \\ &< \beta \end{aligned} \tag{3.8}$$

by the choice of r. It follows that the $x_k(t)$ are *uniformly bounded* in $[t_o, t_o+r]$. Now let ξ_1 and ξ_2 be two points in $[t_o, t_o+r]$ with $|\xi_1-\xi_2| \le 2^{-k}r$. Thus they are either in the same or in adjacent subintervals. In either case

$$\|x_k(\xi_1)-x_k(\xi_2)\| \le m|\xi_1-\xi_2| \tag{3.9}$$

We can dispense with the restriction $|\xi_1-\xi_2| \le 2^{-k}r$ because if $|\xi_1-\xi_2| > 2^{-k}r$, the interval $[\xi_1, \xi_2]$ can be broken up into subintervals of length $< 2^{-k}r$ and (3.9) can be used for each subinterval. The resulting inequalities add to give a sum $\ge \|x_k(\xi_1)-x_k(\xi_2)\|$ on the left, while on the right we still get

$m|\xi_1-\xi_2|$. Hence, for any choice of ξ_1,ξ_2 in $[t_o,t_o+r]$ and for any choice of k, (3.9) holds. Therefore the functions $x_k(t)$ satisfy a fixed *Lipschitz condition* and form an *equicontinuous* family.

Applying the Arzelà-Ascoli theorem **1.2**, we can find a subsequence $\{x_{k_i}(t)\}$ which converges uniformly in $[t_o,t_o+r]$ to a continuous function x(t).

There remains the question of the differentiability of x(t). At the partition points t_{jk} the derivative does not exist, but we have left and right-hand derivatives (Rosenbrock & Storey, *op. cit.*) and for large values of k these differ very little. For an $\xi \in [t_o,t_o+r)$

$$\mathscr{D}^+ x_k(\xi) = f(\xi,x_k(\xi)) \Rightarrow \mathscr{D}^+ x(\xi) = f(\xi,x(\xi)) \tag{3.10}$$

for $\xi = t_o$

Suppose now that $\xi > t_o$ is such that

$$t_{j-1,k_i} < \xi \le t_{j,k_i}$$

Then we have

$$\mathscr{D}^- x_{k_i}(\xi) = f(t_{j-1,k_i}, x_{k_i}(t_{j-1,k_i})) \tag{3.11}$$

$$\mathscr{D}^+ x_{k_i}(\xi) = \begin{cases} \mathscr{D}^- x_{k_i}(\xi) & \text{if } \xi \neq t_{j,k_i} \\ f(t_{j,k_i}, x_{k_i}(t_{j,k_i})) & \text{if } \xi = t_{j,k_i} \end{cases} \tag{3.12}$$

Now

$$\begin{aligned} \|x_{k_i}(t_{j-1,k_i})-x(\xi)\| &\le \|x_{k_i}(t_{j-1,k_i})-x_{k_i}(\xi)\| \\ &\quad + \|x_{k_i}(\xi)-x(\xi)\| \\ &\le m2^{-k_i}r + \|x_{k_i}(\xi)-x(\xi)\| \\ &\to 0 \quad \text{as} \quad i \to \infty. \end{aligned} \tag{3.13}$$

This implies that

$$f(t_{j-1,k_i}, x_{k_i}(t_{j-1,k_i})) \rightarrow f(\xi, x(\xi)) \tag{3.14}$$

since f is continuous in D. Thus $\dot{x}(t)$ exists for all $t \in [t_o, t_o+r]$ and satisfies (3.1).

It is interesting to note that any limit function of the sequence $\{x_k(t)\}$ is a solution to the initial value problem. Thus if the sequence does not have a unique limit, the solution is not unique. Clearly the conditions on f are not strong enough to guarantee uniqueness and additional conditions are required. A convenient condition on f is that it satisfies a *local Lipschitz condition*, which is defined as follows:

Definition 3.1: *Local Lipschitz condition*

A function f(t,x) defined on a domain $D \subset R^{n+1}$ is said to satisfy a local Lipschitz condition in x if for any compact set $U \subset D$, there is a z such that

$$\|f(t,x)-f(t,y)\| \leq z\|x-y\| \tag{3.15}$$

for $(x,t),(y,t) \subset U$.

Note that if f(t,x) has continuous first partial derivatives with respect to x in D, then f(t,x) is locally Lipschitzian in x.

A basic existence and uniqueness theorem under the hypothesis that f(t,x) is locally Lipschitzian in x, is derived from the *Method of Successive Approximations*. This technique is usually attributed to Picard, and leads to the following:

Theorem 3.2: *Existence and Uniqueness (Picard)*

Suppose that $f: D \rightarrow R^n$ is defined and continuous in $D = \{(x, t): \|x-x_o\| < \beta,\ x \in R^n;\ |t-t_o| < \alpha,\ t \in R\}$ and suppose that

$$\|f(t,x)\| \leq m$$

$$\|f(t,x)-f(t,y)\| \leq z\|x-y\|$$

there. Then there is a unique solution of (3.1) passing through (x_o, t_o) which is defined on the interval (t_o-r, t_o+r) where

$r < \min(\alpha, \beta m^{-1})$

Proof is adapted from Hille *(op.cit.)*. The initial value problem (3.2) is replaced by the equivalent integral equation

$$x(t) = x_o + \int_{t_o}^{t} f(z,x(s))ds \tag{3.16}$$

whose solution will be obtained as the limit of a sequence of functions $\{x_k(t)\}$ defined by

$$x_o(t) = x_o$$

$$x_k(t) = x_o + \int_{t_o}^{t} f(s,x_{k-1}(s))ds \tag{3.17}$$

for $k = 1,2,3,\ldots$

Assume that $x_{k-1}(t)$ is well defined on the interval of interest. Clearly, $x_{k-1}(t_o) = x_o$, but the induction hypothesis must also include that $x_{k-1}(t)$ is continuous and $\|x_{k-1}(t)-x_o\| < \beta$. Thus we see that $f(s,x_{k-1}(s))$ is well defined and continuous. Also

$$\|f(s,x_{k-1}(s))\| \leq m$$

whence

$$\begin{aligned} \left\|\int_{t_o}^{t} f(s,x_{k-1}(s))ds\right\| &\leq \int_{t_o}^{t} \|f(s,x_{k-1}(s))\|ds \\ &\leq m|t-t_o| \\ &< mr < b \end{aligned} \tag{3.18}$$

by the choice of r. This implies that $x_k(t)$ is also continuous, and satisfies $x_k(t_o) = x_o$, $\|x_k(t)-x_o\| < \beta$. Thus the approximations are well defined for all k.

Using the assumption that f(t,x) satisfies a local Lipschitz condition, we have

$$\|x_k(t)-x_{k-1}(t)\| = \left\| \int_{t_o}^{t} [f(s,x_{k-1}(s))-f(s,x_{k-2}(s))]ds \right\|$$

$$\leq z \int_{t_o}^{t} \|x_{k-1}(s)-x_{k-2}(s)\|ds \tag{3.19}$$

For $k = 1$ we have the estimate

$$\|x_1(t)-x_o(t)\| = \left\| \int_{t_o}^{t} f(s,x_o(s))ds \right\| \leq zm|t-t_o| \quad \text{for } |t-t_o| < r \tag{3.20}$$

Proceeding by induction, we assume that

$$\|x_{k-1}(t)-x_{k-2}(t)\| \leq \frac{z^{k-2}}{(k-1)!} m|t-t_o|^{k-1} \quad \text{for } |t-t_o| < r \tag{3.21}$$

and infer that

$$\|x_k(t)-x_{k-1}(t)\| \leq \frac{z^{k-1}}{k!} m|t-t_o|^k \tag{3.22}$$

so the estimate holds for all k.

It follows that the series

$$x_o(t) + \sum_{k=1}^{\infty} [x_k(t)-x_{k-1}(t)] \tag{3.23}$$

whose k^{th} partial sum is $x_k(t)$, converges uniformly for $|t-t_o| < r$ to a continuous limit function $x:R \to R^n$. From (3.22) we note that

$$\|x(t)-x_k(t)\| \leq \frac{z^{k-1}}{k!} m \exp(z|t-t_o|)|t-t_o|^k \tag{3.24}$$

and observe that if $|t-t_o|$ is not large, $x_k(t)$ converges rapidly to the limit $x(t)$.

The uniform convergence of $x_k(t)$ to $x(t)$ implies the uniform convergence of $f(t,x_{k-1}(t))$ to $f(t,x(t))$, and

$$\int_{t_o}^{t} f(s,x_{k-1}(s))ds \rightarrow \int_{t_o}^{t} f(s,x(s))ds$$

uniformly in t. Hence x(t) satisfies the integral equation (3.3) and consequently is a solution to the initial value problem (3.2).

To show uniqueness, suppose that there are two solutions $x_1(t)$, $x_2(t)$ defined in the interval (t_o-r,t_o+r). Then we have

$$||x_1(t)-x_2(t)|| = \int_{t_o}^{t} [f(s,x_1(s))-f(s,x_2(s))]ds$$

$$\text{for } |t-t_o| < r$$

Hence

$$||x_1(t)-x_2(t)|| \leq z \int_{t_o}^{t} ||x_1(s)-x_2(s)||ds \tag{3.25}$$

by the Lipschitz condition. To complete the proof of uniqueness requires the application of the following lemma:

Lemma 3.1: *Gronwall-Bellman (Bellman, 1953)*

Let g, f and x be continuous functions on $[t_o,t_1]$ to R. If $f(t) \geq 0$ on $[t_o,t_1]$ and x(t) has the property that for $t \in [t_o,t_1]$

$$x(t) \leq g(t) + \int_{t_o}^{t} f(s)x(s)ds \tag{3.26}$$

then on the same interval

$$x(t) \leq g(t) + \int_{t_o}^{t} g(s)f(s)\exp\left(\int_{s}^{t} f(\tau)d\tau\right)ds \tag{3.27}$$

Proof: This lemma is very important in stability studies since it yields an explicit inequality with respect to x from an implicit inequality. The function

$$y(t) = \int_{t_0}^{t} f(s)x(s)ds \tag{3.28}$$

is continuous and continuously differentiable in $t \varepsilon [t_0,t_1]$ and $y(t_0) = 0$. Moreover we can rewrite (3.26) as

$$x(t) \le g(t) + y(t). \tag{3.29}$$

Since

$$\dot{y}(t) = f(t)x(t) \tag{3.30}$$

and

$$f(t)x(t) \le f(t)g(t) + f(t)y(t) \tag{3.31}$$

it follows that

$$\dot{y}(t) \le f(t)g(t) + f(t)y(t) \tag{3.32}$$

Consider now

$$z(t) = y(t)\exp[-\int_{t_0}^{t} f(s)ds] \tag{3.33}$$

Then

$$\begin{aligned} \dot{z}(t) &= \dot{y}(t)\exp[-\int_{t_0}^{t} f(s)ds] - y(t)f(t)\exp[-\int_{t_0}^{t} f(s)ds] \\ &\le f(t)g(t)\exp[-\int_{t_0}^{t} f(s)ds] \end{aligned} \tag{3.34}$$

Obviously $z(t_0) = 0$. Integrating both sides of (3.34) between t_0 and $t \varepsilon [t_0,t_1]$ gives

$$z(t) \le \int_{t_0}^{t} f(s)g(s)\exp[\int_{s}^{t} f(\tau)d\tau]ds \tag{3.35}$$

whence, by (3.33)

$$y(t) \le \int_{t_0}^{t} f(s)g(s)\exp[\int_{s}^{t} f(\tau)d\tau]ds \tag{3.36}$$

Therefore

$$x(t) \leq g(t) + \int_{t_o}^{t} f(s)g(s)\exp[\int_{s}^{t} f(\tau)d\tau]ds$$

which is the required inequality.

To return to the proof of theorem 3.2, we set $x(t) = \|x_1(t) - x_2(t)\|$, $g(t) = 0$ and $f(t) = z$ in (3.26) and deduce that

$$\|x_1(t) - x_2(t)\| \leq 0 \tag{3.37}$$

so that

$$x_1(t) = x_2(t)$$

and uniqueness is proved.

It must be emphasized that the above theorems establish only the *local* existence of solutions to the initial value problem. In general, the continuation of solutions for $|t-t_o| > r$ is difficult and is not considered here. The continuous dependence of solutions on initial data is contained in theorem 3.2 by virtue of the Lipschitz condition satisfied by $f(t,x)$, although the question in general is complicated. However, on infinite time intervals we arrive at the concept known as *stability in the sense of Liapunov*, a topic we will return to in Chapter Five. The important case when $f(t,x)$ is periodic or more generally almost periodic can not be treated in this chapter since the existence conditions assume some form of stability property for the solution to (3.1) and is therefore left until Appendix One.

3.3 Linear Ordinary Differential Equations

In the remaining sections of this chapter we are going to examine in great detail the properties of linear ordinary differential equations. These are of great importance because they are very frequently used to represent the dynamical behaviour of many physical systems encountered in engineering practice. Furthermore, we shall see that it is often possible to give explicit solutions to equations of this type. First of all we recast the existence and uniqueness theorem (theorem **3.2**) for linear equations. The crucial point is that the solution of a linear ordinary differential equation is defined wherever the equation is defined. If we denote M_n as the set of n×n bounded and continuous real or complex valued matrix functions of the real variable t, the existence and uniqueness condition for the linear case is given by:

Theorem 3.3: *Existence and Uniqueness - Linear Case*

Suppose that $A:J \to M_n$ is defined and continuous in $J = \{t: |t-t_o| < \alpha, \ t \in R\}$ and suppose that $(x_o,t_o) \in R^n \times J$. Then there is a unique solution of

$$\dot{x}(t) = A(t)x(t) \tag{3.38}$$

passing through (x_o,t_o) which is defined on the whole interval $|t-t_o| < \alpha \leq \infty$.

Proof: As in the proof of theorem **3.2** the method of successive approximations can be used. The details are given in Brockett (1970). Note that on compact intervals $||A|| = \sum_{i,j} \max_t |a_{ij}(t)|$ n^2N if $|a_{ij}| \leq N$ in the interval of interest. Hence the Lipschitz condition

$$||A(t)x - A(t)y|| \leq z||x-y|| \tag{3.39}$$

is clearly satisfied with $z = n^2N$ and guarantees uniqueness.

The next step, having established existence and uniqueness for the linear case, is to consider the problem of constructing explicit solutions to the initial value problem. The technique is contained in the following:

Theorem 3.4: *Solution space*

Let S denote the set of all solutions of equation (3.38), i.e. of the equation

$$\dot{x} = A(t)x$$

with A defined and continuous as above. In other words,

$$S = \{x(t): \dot{x}(t) = A(t)x(t), \quad t \in J\} \tag{3.40}$$

Then S is an n-dimensional vector space, and a basis $\{x_1(t), x_2(t),\ldots,x_n(t)\}$ of S may be obtained by letting $x_i(t)$ be the *unique* element of S which satisfies the condition

$$x_i(t_o) = e_i \tag{3.41}$$

where $\{e_1,e_2,e_3,\ldots,e_n\}$ is the natural basis of R^n.

Proof: Clearly the function $x(t) = 0$ for all $t \in J$ is in S. This function is called the *trivial* solution. Furthermore, if $\zeta,\xi \in R$ and $x(t),y(t) \in S$, then

$$\begin{aligned} \frac{d}{dt}[\zeta x(t) + \xi y(t)] &= \zeta\dot{x}(t) + \xi\dot{y}(t) \\ &= \zeta A(t)x(t) + \xi A(t)y(t) \\ &= A(t)[\zeta x(t) + \xi y(t)] \end{aligned} \tag{3.42}$$

so that $[\zeta x(t) + \xi y(t)] \in S$. It is a simple matter to verify that S forms an additive group under addition and that scalar multiplication has the following properties:

$$\begin{aligned} &\text{(i)} \quad \zeta[x(t)+y(t)] = \zeta x(t) + \zeta y(t) \\ &\text{(ii)} \quad (\zeta+\xi)x(t) = \zeta x(t) + \xi x(t) \\ &\text{(iii)} \quad \zeta[\xi x(t)] = [\zeta\xi]x(t) \\ &\text{(iv)} \quad 1x(t) = x(t) \end{aligned} \tag{3.43}$$

where $\zeta,\xi \in R$ and 1 is the identity of R.

We now show that the $x_i(t)$ are a basis of S. First of all suppose that

$$\sum_{i=1}^{n} c_i x_i(t) = 0 \qquad \forall t \in J \tag{3.44}$$

Then

$$\sum_{i=1}^{n} c_i x_i(t_o) = \sum_{i=1}^{n} c_i e_i = 0 \tag{3.45}$$

and so the scalars c_i are all 0. This implies that the $x_i(t)$ are linearly independent functions. Now, if x(t) is any element of S then $x(t_o)$ can be constructed by taking linear combinations of the e_i, that is

$$x(t_o) = \sum_{i=1}^{n} \theta_i e_i \tag{3.46}$$

The function $\sum_{i=1}^{n} \theta_i x_i(t)$ is in S and agrees with x(t) at t_o by (3.46). Hence it follows from the uniquess part of theorem 3.3 that

$$x(t) = \sum_{i=1}^{n} \theta_i x_i(t) \qquad \forall t \in J \tag{3.47}$$

and the assertion is proved.

This theorem leads us to the important definition (Yakubovich and Starzhinskii, (1975)):

Definition 3.2: *Fundamental matrix*

Let $\Phi(t,t_o)$ be the n×n matrix whose j^{th} column is the vector $x_j(t) \in S$, with $x_j(t_o) = e_j$. That is to say, the columns of $\Phi(t,t_o)$ are n-linearly independent solutions of (3.38) satisfying the initial condition $x_j(t_o) = e_j$, $j = 1,2,\ldots,n$. Then we call $\Phi(t,t_o)$ the *fundamental* or *transition matrix* of the linear ordinary differential equation (3.38).

Note that according to the above definition, $\Phi(t_o,t_o) = I$, where I is the identity matrix. It is clear from the definition of the fundamental matrix that it may be considered as the unique solution to the initial value problem

$$\dot{\Phi} = A(t)\Phi, \quad \Phi(t_o,t_o) = I, \quad t \in J \tag{3.48}$$

Moreover, if $(x_o,t_o) \in R^n \times J$, it follows that the solution of equation (3.38) passing through (x_o,t_o) is given by

$$x(t) = \Phi(t,t_o)x_o \tag{3.49}$$

Next we aim to show that the matrix $\Phi(t,t_o)$ is *nonsingular* for all $t \in J$. Use will be made of the following lemma whose origin is attributed to Abel, Jacobi and Liouville.

Lemma 3.2: *Abel-Jacobi-Liouville*

Suppose that $\Phi(t,t_o)$ is a fundamental matrix for $\dot{x} = A(t)x$, $A:J \to M_n$ is defined and continuous in J. Then

$$\det \Phi(t,t_o) = \exp\left[\int_{t_o}^{t} \operatorname{tr} A(s)ds\right] \tag{3.50}$$

for all $t, t_o \in J$.

Proof: Brockett *(op.cit.)* demonstrates this lemma by showing that $\det\Phi(t,t_o)$ is a solution to the *scalar* initial value problem $\dot{w} = [\operatorname{tr} A(t)]w$, $w_o = 1$, $t_o \in J$. This lemma is important in stability studies for non-autonomous systems since it is clear that the linear system (3.38) is *unstable* if $\lim_{t\to\infty} \operatorname{Re}\left\{\int_{t_o}^{t} \operatorname{tr}A(\tau)d\tau\right\} = +\infty$, since $\det\Phi(t,t_o)$ is unbounded.

Theorem 3.5

Suppose that $\Phi(t,t_o)$ is a fundamental matrix for $\dot{x} = A(t)x$, where $A:J \to M_n$ is defined and continuous in J. Then $\Phi(t,t_o)$ is nonsingular for all $t \in J$.

Proof: follows immediately from lemma 3.2, i.e. $\det\Phi(t,t_o) \neq 0$ for all $t \in J$.

Having established that $\Phi(t,t_o)$ is a nonsingular matrix, it is of interest to try to find its inverse. As a consequence of the definition of $\Phi(t,t_o)$, which includes uniqueness, it is intuitively clear that fundamental matrices satisfy the following *composition rule:*

$$\Phi(t,t_o) = \Phi(t,t_1)\Phi(t_1,t_o) \tag{3.51}$$

$$\text{for} \quad t, t_1, t_o \in J.$$

Since $\Phi(t_o,t_o) = I$, it follows that

$$\Phi^{-1}(t_1,t_o) = \Phi(t_o,t_1) \tag{3.52}$$

For arbitrary $t = t_1 \in J$ we have $\Phi^{-1}(t,t_o) = \Phi(t_o,t)$ by the same argument.

Lemma 3.3

Suppose that $\Phi(t,t_o)$ is a fundamental matrix for $\dot{x} = A(t)x$, where $A:J \to M_n$ is defined and continuous in J. Then the matrix $\Phi^{-1}(t,t_o) = \Phi(t_o,t)$ is a fundamental matrix of the *adjoint* differential equation

$$\dot{y} = -yA(t) \tag{3.53}$$

where $y':J \to R^n$, i.e. y is a *row* vector, is continuous and differentiable in J.

Proof: Since $\Phi(t_o,t)\Phi(t,t_o) = I$ it follows that

$$0 = \dot{\Phi}(t_o,t)\Phi(t,t_o) + \Phi(t_o,t)\dot{\Phi}(t,t_o)$$

$$= \dot{\Phi}(t_o,t)\Phi(t,t_o) + \Phi(t_o,t)A(t)\Phi(t,t_o)$$

Hence, noting that $\det\Phi(t,t_o) \neq 0$

$$\dot{\Phi}(t_o,t) = -\Phi(t_o,t)A(t) \tag{3.54}$$

Recalling the definition of fundamental matrices, equation (3.54) is equivalent to saying that each row of the matrix $\Phi(t_o,t)$ is a solution of (3.53).

A simple transposition of equation (3.54) yields

$$\dot{\Phi}'(t_o,t) = -A'(t)\Phi'(t_o,t) \tag{3.55}$$

If we have $A(t) = -A'(t)$ then equation (3.53) is called *self adjoint.* Such equations are found in the study of mechanics, often in connection with oscillator problems (Venkatesh, 1977). The fundamental matrix of any self-adjoint equation is clearly *orthogonal.* This implies that every solution of (3.53) has a constant norm as t varies.

The final topic we include in this section concerns the addition of a forcing term to the initial value problem (3.38), that is to say we now have an inhomogeneous linear differential equation of the form

$$\dot{x}(t) = A(t)x(t) + h(t) \tag{3.56}$$

where $A:J \to M_n$ is defined and continuous in J, $x:J \to R^n$ is continuous and differentiable in J and $h:J \to R^n$ is continuous in J, which satisfies the boundary condition $(x_o,t_o) \in R^n \times J$. To construct a solution to (3.56) we appeal to the following:

Theorem 3.6: *Variation of constants*

Suppose that $\Phi(t,t_o)$ is a fundamental matrix for (3.38). Then every solution of (3.56) is given by the formula

$$x(t) = \Phi(t,t_o)x_o + \int_{t_o}^{t} \Phi(t,s)h(s)ds \tag{3.57}$$

for any $t,t_o \in J$, $x_o \in R^n$.

Proof: If we rewrite equation (3.56) as $\dot{x}-A(t)x = h$ and use (3.54), then equation (3.56) is equivalent to

$$\frac{d}{ds}[\Phi^{-1}(s,t_o)x(s)] = \Phi^{-1}(s,t_o)h(s) \tag{3.58}$$

Integrating both sides from t_o to t we obtain

$$\Phi^{-1}(t,t_o)x(t) - \Phi^{-1}(t_o,t_o)x_o = \int_{t_o}^{t} \Phi^{-1}(s,t_o)h(s)ds$$

A rearrangement of the terms in this expression yields equation (3.57) and the theorem is proved.

3.4 Constant Coefficient Differential Equations

We shall now turn our attention to the special, but very important, case of linear ordinary differential equations with constant coefficients. In other words, we shall examine initial value problems of the form:

$$\dot{x}(t) = Ax(t), \quad x(t_o) = x_o, \quad t_o \in J \tag{3.59}$$

where $A \in \overline{M}_n$, the set of n×n constant coefficient matrices, $x:J \to R^n$ is continuous and differentiable in J. To begin with the fundamental matrix for (3.59) satisfies

$$\dot{\Phi}(t,t_o) = A\Phi(t,t_o), \quad \Phi(t_o,t_o) = I \tag{3.60}$$

for all $t,t_o \in J$. It follows that $\Phi(t,t_o)$ must satisfy the integral equation

$$\Phi(t,t_o) = I + \int_{t_o}^{t} A\Phi(s,t_o)ds \tag{3.61}$$

If we attempt to solve equation (3.61) by the method of successive approximations

$$\Phi_o(t,t_o) = I$$

$$\Phi_{k+1}(t,t_o) = I + \int_{t_o}^{t} A\Phi_k(s,t_o)ds$$

$$k = 0,1,2,\ldots,$$

then we observe that the expression for $\Phi_k(t,t_o)$ is given by the formula

$$\Phi_k(t,t_o) = I + A(t-t_o) + \frac{A^2(t-t_o)^2}{2!} + \frac{A^3(t-t_o)^3}{3!} + \ldots$$

$$\ldots + \frac{A^k(t-t_o)^k}{k!}$$

which we recognise, as $k \to \infty$, as the power series expansion of the exponential matrix function. By the ratio test (Rosenbrock & Storey, *op.cit.*) this series converges for all $|t-t_o| < \infty$. From this we deduce that $J = R$ and

$$\Phi(t,t_o) = \sum_{k=0}^{\infty} \frac{A^k(t-t_o)^k}{k!}$$

$$= \exp[A(t-t_o)], \quad \forall t,t_o \in R \tag{3.62}$$

so that all solutions to the initial value problem (3.59) can be expressed very succinctly as

$$x(t) = \exp[A(t-t_o)]x_o \tag{3.63}$$

For convenience we recall some basic properties of matrix exponentials:

(i) $\exp[At]\exp[As] = \exp[A(t+s)]$

$$\text{(ii)}\quad (\exp[At])^{-1} = \exp[-At]$$

$$\text{(iii)}\quad \exp[A]\exp[B] = \exp[A+B] \quad \text{iff } AB = BA \tag{3.64}$$

$$\text{(iv)}\quad \frac{d}{dt}\exp[At] = A\exp[At] = \exp[At]A$$

$$\text{(v)}\quad \exp[P^{-1}AP] = P^{-1}\exp[A]P \quad \text{if } \det P \neq 0$$

where $A,B,P \in \overline{M}_n$ and $t,s \in R$.

It is of interest to see the explicit form that the fundamental matrix $\Phi(t,t_o)$ assumes in relation to the constant matrix A. From the theory of matrices (see for example Gantmacher, 1959) it is known that there exists a nonsingular constant matrix $P \in \overline{M}_n$ such that $P^{-1}AP$ is the Jordan canonical form of A, i.e.,

$$P^{-1}AP = A_J \tag{3.65}$$

Then

$$\begin{aligned}\Phi(t,0) &= \exp[At] \\ &= \exp[PA_JP^{-1}t] \\ &= P\exp[A_Jt]P^{-1}\end{aligned} \tag{3.66}$$

and A_J has the form

$$A_J = \operatorname{diag}[A_1,A_2,\ldots,A_k] \tag{3.67}$$

where each square block

$$A_j = \operatorname{diag}[A_{j1},A_{j2},\ldots,A_{jm_j}], \quad j=1,2,\ldots,k \tag{3.68}$$

and each square elementary divisor block

$$A_{jp} = \begin{bmatrix} \lambda_j & 1 & \cdot & 0 \\ & \lambda_j & 1 & \cdot \\ & & \cdot & 1 \\ 0 & \cdot & \cdot & \lambda_j \end{bmatrix}, \quad p=1,2,\ldots,m_j \tag{3.69}$$

The latter has dimension r_{pj}. Using this notation for A_J,

$$\sum_{j=1}^{k}\sum_{p=1}^{m_j} r_{pj} = n \tag{3.70}$$

We see that in general A_J will have k distinct *characteristic values* or *eigenvalues* λ_j, each being associated with m_j elementary divisor blocks of dimension r_{pj}. To within the ordering of the elementary divisor blocks, this representation is unique.

Theorem 3.7

Suppose that $\Phi(t,0)$ is a fundamental matrix for $\dot{x}(t) = A_J x(t)$ where $x:R \to R^n$ is continuous and differentiable in R. Then the columns of $\Phi(t,0) = \exp[A_J t]$ are given by

$$y(t)_{j,p,q} = \exp(\lambda_j t) \sum_{h=0}^{q-1} \frac{t^h}{h!} e_{j,p,q-h} \tag{3.71}$$

for $q=1,2,\ldots,r_{pj}$, $p=1,2,\ldots,m_j$, $j=1,2,\ldots,k$. The terms $e_{j,p.q-h}$ are columns of the identity matrix.

Proof: Coddington & Levinson *(op.cit.)* and Gantmacher *(op.cit.)* give proofs for a general matrix $A \in \overline{M}_n$.

To discover a little more about the structure of the fundamental matrix $\exp[A_J t]$ we rewrite (3.71) as

$$y(t)_{j,p,q} = \exp(\lambda_j t)t^{q-1} \left(\frac{1}{(q-1)!} e_{j,p,1} + \sum_{h=0}^{q-2} \frac{t^{h+1-q}}{h!} \times e_{j,p,q-h} \right) \tag{3.72}$$

and note that the term after the summation sign in the brackets tends to zero as $t \to \infty$ although the norm of the brackets remains bounded away from zero. Therefore we have

$$\log\|y(t)_{j,p,q}\| = \chi_j t + (q-1)\log t + \log\|[\cdot]\| \tag{3.73}$$

where $[\cdot]$ is the bracketed expression in (3.72). From (3.73) it follows that

$$\chi_j = \overline{\lim_{t\to+\infty}} \left\{ t^{-1} \log\|y(t)_{j,p,q}\| \right\} \tag{3.74}$$

and

$$q-1 = \overline{\lim_{t\to+\infty}} \left\{ \frac{\log\|y(t)_{j,p,q} \exp(-\chi_j t)\|}{\log t} \right\} \tag{3.75}$$

where

$$\chi_j = \operatorname{Re}(\lambda_j) \tag{3.76}$$

It follows that the multiplicity of χ_j, which will be denoted by $\mu(\chi_j)$, is given by

$$\mu(\chi_j) = \sum_{p=1}^{m_j} r_{pj} \tag{3.77}$$

since the matrix A_J possesses k distinct characteristic values. The relationship (3.76) between χ_j and λ_j leads to the following definition, originally due to Liapunov (1893) and generalised by Perron (1930):

Definition 3.3: *Generalised Characteristic Exponents*

Suppose that $\dot{x}(t) = A(t)x(t)$, where $A:R \to M_n$ is defined and continuous in R, $x:R \to R^n$ is continuous and differentiable in R. For each nontrivial solution x(t) define a *generalised characteristic exponent* $\chi(x)$ of A(t) by

$$\chi(x) = \overline{\lim_{t\to+\infty}} \left\{ t^{-1} \log\|x(t)\| \right\} \tag{3.78}$$

Furthermore, if $A(t) = A \in \overline{M}_n$, then the χ_j are precisely the real parts of the characteristic values λ_j, $j=1,2,\ldots,k$ of A.

Generalised characteristic exponents play an important role in stability theory, as we shall see in the next section and in Chapter Five. A related concept is the *type number* of a matrix, which was introduced by Marcus (1955):

Definition 3.4: *Type Number*

Suppose that $\dot{x}(t) = A(t)x(t)$, where $A:R \to M_n$ is defined and continuous in R, $x:R \to R^n$ is continuous and differentiable in R. For each nontrivial solution x(t) define a *type number* $\nu(\chi)$ of A(t) by

$$\nu(\chi) = \overline{\lim_{t\to+\infty}} \left\{ \frac{\log\|x(t)\exp(-\chi t)\|}{\log t} \right\} \tag{3.79}$$

Each χ_j may have many type numbers associated with it and they are not necessarily distinct. We use the notation ν_{jw} to represent distinct type numbers and let $\mu(\nu_{jw})$ denote their multiplicity. It is interesting to observe that in the case $A(t) = A_J \varepsilon \overline{M}_n$, the dimensions of the blocks A_{jp}, $p=1,2,\ldots,m_j$ are determined by the type numbers ν_{jw}. In fact, the number of blocks A_{jp} of dimension $\geq \nu_{jw}+1$ is exactly $\mu(\nu_{jw})$. Thus the total structure of A_J is determined to within the imaginary parts of the characteristic values of A by the χ_j, ν_{jw}, $\mu(\chi_j)$, $\mu(\nu_{jw})$.

3.5 Periodic Coefficients and Floquet Theory

Linear ordinary differential equations with periodic coefficients occur in many theoretical and practical problems concerned with rotational or vibrational motion (Stephens, 1966). Herein lies their importance. We consider the following initial value problem

$$\dot{x}(t) = A(t)x(t), \quad A(t+w) = A(t), \quad (x_o,t_o) \varepsilon R^{n+1} \tag{3.80}$$

where $A:R \rightarrow AP_n$ and $x:R \rightarrow R^n$ is continuous and differentiable in R. We note that if $\Phi(t,t_o)$ is a fundamental matrix of (3.80), then so is $\Phi(t+w,t_o)$. Thus we may write

$$\begin{aligned} \Phi(t+w,0) &= \Phi(t+w,w)\Phi(w,0) \\ &= \Phi(t,0)\Phi(w,0) \end{aligned} \tag{3.81}$$

Since $\Phi(w,0)$ is nondegenerate, there exists a matrix which will be denoted by $\log\Phi(w,0)$, such that

$$\exp[\log\Phi(w,0)] = \Phi(w,0) \tag{3.82}$$

This assertion has been demonstrated by Bellman *(op.cit.)* for example. This leads to the following:

Theorem 3.8: *Floquet Representation (Floquet, 1883)*

The fundamental matrix of (3.80) can be represented in the form

$$\Phi(t,0) = P(t)\exp[w^{-1}t \log\Phi(w,0)] \tag{3.83}$$

where $P(t)$ is a periodic nonsingular matrix and $P(0) = I$.

Proof: We write

$$P(t) = \Phi(t,0)\exp[-w^{-1}t \log\Phi(w,0)]$$

Then P(t) is nonsingular, and we have

$$\begin{aligned} P(t+w) &= \Phi(t+w,0)\exp[-w^{-1}(t+w)\log\Phi(w,0)] \\ &= \Phi(t,0)\Phi(w,0)\exp[-\log\Phi(w,0)]\exp[-w^{-1}t \log\Phi(w,0)] \\ &= P(t) \end{aligned} \tag{3.84}$$

For general $t_0 \neq 0$ it follows that

$$\Phi(t,t_0) = P(t)\exp[w^{-1}(t-t_0)\log\Phi(w,0)]P^{-1}(t_0) \tag{3.85}$$

Theorem 3.9

There exists a nonsingular periodic transformation of variables which transforms (3.80) into an equation with constant coefficients.

Proof: Suppose that $P(t) = P(t+w)$ is defined by (3.83). To simplify notation, let

$$B = w^{-1}\log\Phi(w,0) \tag{3.86}$$

Let $x(t) = P(t)y(t)$ in (3.80). The equation for y is

$$\dot{y} = P^{-1}(t)[A(t)P(t)-\dot{P}(t)]y \tag{3.87}$$

Since $P(t) = \Phi(t,0)\exp[-Bt]$, it follows that

$$B = P^{-1}(t)[A(t)P(t)-\dot{P}(t)] \tag{3.88}$$

and this proves the result.

Note that the characteristic values of $\Phi(w,0)$ are generally complex numbers and that their logarithms are also complex, although the real part is determined uniquely. There are occasions when A(t), P(t) and B are real, even though the characteristic values of $\Phi(w,0)$ may be negative. In such cases the matrix P(t) has period 2w and is of the form

$$P(t) = \Phi(t,0)\exp[-(2w)^{-1}t(\log\Phi(w,0) + \log^*\Phi(w,0))]$$

where the asterisk denotes complex conjugate.

Definition 3.5: *Characteristic Multiplier*

Suppose that $\Phi(t,0)$ is the fundamental matrix of (3.80). Then the characteristic values of $\Phi(w,0)$ are called *characteristic multipliers* of (3.80) and are denoted by ρ_j, $j=1,2,\dots,n$.

Observe that $Re(\lambda)$ such that $\rho = \exp[\lambda w]$ is a generalised characteristic exponent of the periodic matrix $A(t)$. We now have the following:

Theorem 3.10

If $\rho_j = e^{\lambda_j w}$, $j=1,2,\dots,n$ are the characteristic multipliers of (3.80), then

$$\prod_{j=1}^{n} \rho_j = \exp\left(\int_0^w trA(s)ds\right)$$

$$\sum_{j=1}^{n} \lambda_j = \frac{1}{w}\int_0^w trA(s)ds, \quad \left(\mod \frac{2\pi i}{w}\right)$$

Proof: The theorem follows immediately from the definitions of characteristic multipliers and exponents, and Lemma 3.2.

It would appear that linear periodic equations share the same simplicity as linear equations with constant coefficients. However, there is a very important difference - the characteristic exponents are defined explicitly only after the solutions of (3.80) are known, there being no obvious relation between the characteristic exponents and the matrix $A(t)$.

Clearly the problem of determining the characteristic multipliers or exponents of linear periodic differential equations is an extremely difficult one. In fact it has no known solution except for some scalar second order systems and, more generally, Hamiltonian and canonical systems. The problem is further examined in subsequent chapters.

3.6 Notes

Linear periodic differential systems have attracted considerable interest (Starzhinskii (1955), Cesari (1963), Yakubovich and Starzhinskii (1975), Venkatesh (1977)); an important special case is the scalar second order differential equation *(Hill's equation)*,

$$\ddot{x} + f(t)x = 0, \qquad \text{with} \quad f(t+w) = f(t) \tag{3.89}$$

Liapunov (1949) has shown that if $f(t) \le 0, \forall t$, then the characteristic multipliers of Hill's equation are real and positive so that there are infinitely many unbounded solutions $\{x(t)\}$ for $t \in R_+$. However if $f(t) > 0, \forall t$ and $\frac{1}{w}\int_0^w f(\tau)d\tau \le 4$, then Hill's equation has purely complex characteristic multipliers with all solutions bounded for $t \in R_+$. Related to Hill's equation is the well known *Mathieu* equation, in which $f(t) = \delta + \xi\cos 2t$ with $w = \pi$ and the *Meissner* equation in which

$$f(t) = \delta + \xi a(t), \qquad \text{for} \quad a(t) = \begin{cases} 1 & 0 \le t < \pi \\ -1 & \pi \le t \le 2\pi \end{cases}$$

and $w = 2\pi$; for further references on these special periodic equations see Cesari (1963) and McLachlean (1947).

3.7 References

Birkhoff, G. and Rota, G. (1978). "Ordinary differential equations", J. Wiley, New York

Bellman, R. (1953). "Stability theory of differential equations", McGraw-Hill, New York

Brockett, R.W. (1970). "Finite dimensional linear systems", J. Wiley, New York

Cesari, L. (1963). "Asymptotic behaviour and stability problems in ordinary differential equations", 2nd Ed. Academic Press, New York

Coddington, E.A. and Levinson, N. (1955). "Theory of ordinary differential equations", McGraw-Hill, New York

Curtain, R.F. and Pritchard, A.J. (1977). "Functional analysis in modern applied mathematics", Academic Press, New York

Floquet, G. (1893). *Ann.Ecole Norm.Sup.* **12**, 47-79

Gantmacher, F.R. (1959). "The theory of matrices", Chelsea, New York

Hale, J.K. (1969). "Ordinary differential equations", Wiley Interscience, New York

Hille, E. (1969). "Lectures on ordinary differential equations", Addison-Wesley, New Jersey

Liapunov, A.M. (1893). *Comm.Soc.Math.Kharkov* (in Russian). (In translation: *Ann.Math.Studies* **17**, Princeton, (1949))
Markus, L. (1955). *Math.Zeits.* **62**, 310-319
McLachean, N.W. (1947). "Theory and applications of Mathieu functions", Clarendon Press, Oxford
Perron, O. (1930). *Math.Zeits* **32**, 703-728
Rosenbrock, H.H. and Storey, C. (1970). "Mathematics of dynamical systems", Nelson, London
Starzhinskii, V.M. (1955). *Amer.Math.Soc.Trans.* **1**, 189
Stevens, K.K. (1966). *SIAM J.Appl.Math.* **14**, 782
Venkatesh, Y.V. (1977). "Energy methods in time-varying system stability and instability analyses", Lecture notes in physics **No.68**, Springer Verlag, Berlin
Yakubovich, V.A. and Starzhinskii, V.M. (1975). "Linear differential equations with periodic coefficients", J. Wiley, New York

Chapter 4

KINEMATIC SIMILARITY

4.1 Introduction

The concept of *kinematic similarity* of matrices as considered by Markus (1955) and others, defines an equivalence relation on the set of all $n \times n$ matrices whose entries are continuous and bounded functions on the nonnegative reals R_+. Evidently this idea has its origins in Liapunov's work (Liapunov, 1893) where it is referred to as *reducibility*, a term which still persists in more recent Soviet literature, e.g. Erugin, 1946. Markus' concept is too restrictive for studies concerned with almost periodic matrices (Langenhop, 1960). Accordingly it is useful to redefine the set of matrices of interest to account for functions which are bounded and continuous on the whole real line R. The resulting modification of Markus' concept Langenhop *(op.cit.)* is called *complete kinematic similarity*.

The purpose of this chapter is threefold. The first is to give a full account of kinematic similarity, roughly following Markus' work. The question of invariants under kinematic similarity is raised and conditions for kinematic similarity stated. Proofs are only given in cases of special interest or significance; in all others the proofs may be found in the cited references. The second objective is to examine the consequences of kinematic similarity in the study of linear systems of ordinary differential equations. The result of great practical significance here is that for linear systems *uniform stability* (see Chapter **5** for a detailed explanation of this concept) is invariant under kinematic similarity. In the last section we give several examples of kinematically similar matrices, each example arising from particular properties of different subsets of M_n.

4.2 Liapunov Transformations and Kinematic Similarity

An important subset of M_n consists of nonsingular matrices whose derivative with respect to time is continuous and bounded on R. An analogous subset is defined on R_+. The matrices referred to are called *Generalised Liapunov Transformations* or simply *Liapunov Transformations* if the range of interest is R_+.

Definition 4.1: *Generalised Liapunov Transformation*

A matrix $P(t) \in M_n$ is said to be a *Generalised Liapunov Transformation* (GLT) if and only if:

(i) $\dot{P}(t) \in M_n$ for all $t \in R$

(ii) There exists some constant $\delta \in R$ such that

$0 < \delta < |\det P(t)|$ for all $t \in R$.

In some applications, such as those concerned with stability theory, it is more convenient to confine interest to the half-line R_+, in which case we have the following:

Definition 4.2: *Liapunov Transformation*

A matrix $P(t) \in M_{n+}$ is said to be a *Liapunov Transformation* (LT) if and only if:

(i) $\dot{P}(t) \in M_{n+}$ for all $t \in R_+$

(ii) There exists some constant $\delta \in R$ such that

$0 < \delta < |\det P(t)|$ for all $t \in R_+$.

Definition 4.3: *Complete kinematic similarity (Langenhop)*

Let $A(t), B(t) \in M_n$. Then $A(t) \sim B(t)$ (read: A(t) *completely kinematically similar* to B(t)) if and only if there exists a Generalised Liapunov Transformation P(t) such that

$$B(t) = -P(t)^{-1}[\dot{P}(t)-A(t)P(t)] \quad \text{for all } t \in R.$$

It is easy to see that we also have

$$A(t) = -Q(t)^{-1}[\dot{Q}(t)-B(t)Q(t)]$$

where $Q(t) = P(t)^{-1}$. Note that the indicated inverse exists by hypothesis. The modifier "complete" is used to distinguish this concept from that considered by Markus *(op.cit.)*, namely:

Definition 4.4: *Kinematic similarity (Markus)*

Let $A(t), B(t) \in M_{n+}$. Then $A(t) \sim B(t)$ if and only if there exists a LT $P(t)$ such that

$$B(t) = -P(t)^{-1}[\dot{P}(t)-A(t)P(t)] \quad \text{for all} \quad t \in R_+.$$

If in the above definition the relationship between $A(t)$ and $B(t)$ is only established to within an arbitrarily small degree there results the condition known as *approximate similarity*, which was introduced by Lillo (1961) in the study of the stability of perturbed differential equations.

Definition 4.5: *Approximate similarity (Lillo)*

Let $A(t), B(t) \in M_n$. Then $A(t) \simeq B(t)$ (read: $A(t)$ *approximately similar* to $B(t)$) if and only if, given any $\delta > 0$ there exists a matrix $P(\delta,t)$ such that $P(\delta,t)$, $P(\delta,t)^{-1}$, $\dot{P}(\delta,t) \in M_n$ and

$$\|-P(\delta,t)^{-1}[\dot{P}(\delta,t)-A(t)P(\delta,t)]-B(t)\| < \delta$$

for all $t \in R$. Here $\|\cdot\|$ is the uniform norm.

This definition is included for completeness and will not be used in the sequel. From definitions 4.3 and 4.4 it is easy to verify that complete kinematic (c.k.) similarity and kinematic (k) similarity are equivalence relations on M_n and M_{n+} respectively.

Theorem 4.1: *(Martin, 1966; Markus, op.cit.)*

Complete kinematic (respectively kinematic) similarity is an *equivalence relation* on M_n (respectively M_{n+}).

Proof: Consider only the case for M_n. It is necessary and sufficient to establish that c.k. similarity is symmetric, reflexive and transitive. Let $A(t), B(t) \in M_n$. Under the identity matrix

$A(t) \sim A(t)$ and we have already noted that if $A(t) \sim B(t)$ then $B(t) \sim A(t)$. From the equation

$$B(t) = -P(t)^{-1}[\dot{P}(t)-A(t)P(t)]$$

it follows that for any GLT $P(t)$, $B(t) \in M_n$ if $A(t) \in M_n$ thus satisfying the initial hypothesis. Finally, if

$$B(t) = -P_1(t)^{-1}[\dot{P}_1(t)-A(t)P_1(t)]$$

and if $C(t) \in M_n$ (by hypothesis) is given by

$$C(t) = -P_2(t)^{-1}[\dot{P}_2(t)-B(t)P_2(t)]$$

then

$$C(t) = -P_3(t)^{-1}[\dot{P}_3(t)-B(t)P_3(t)]$$

where $P_3(t) = P_1(t)P_2(t)$. Clearly $P_3(t)$ is a GLT, so $A(t) \sim C(t)$.

4.3 Invariants and Canonical Forms

Our principal aim in this section is to clarify the relationship between the characteristic exponents of a matrix and kinematic similarity and to show that real Jordan matrices with an assumed structure (permuted) are canonical forms for a subset of M_{n+}, including constant matrices in $\overline{M}_n$. First of all we present the following theorem concerning the invariants of matrices under kinematic similarity and then go on to remark that these form a complete set of invariants for real Jordan matrices.

Theorem 4.2: *Invariants of matrices under kinematic similarity (Markus, 1955)*

The characteristic exponents χ_j of $A(t) \in M_{n+}$, their types ν_{jw} and the multiplicities $\mu(\chi_j)$ and $\mu(\nu_{jw})$ are invariants of kinematic similarity.

Proof: Let

$$-P(t)^{-1}[\dot{P}(t)-A(t)P(t)] = B(t),$$

define the kinematic similarity $A(t) \sim B(t)$, and consider a vector $x:R_+ \to R^n$ satisfying $\dot{x} = A(t)x$ on the half-line R_+

and a vector $y:R_+ \to R^n$ satisfying $\dot{y} = B(t)y$. Then it is easily shown that $x(t) = P(t)y(t)$. Consequently

$$\chi(x) = \chi(P(t)y(t)) = \overline{\lim_{t\to+\infty}}\, t^{-1}\log\|P(t)y(t)\|$$

$$\leq \overline{\lim_{t\to+\infty}}\, t^{-1}\log\|P(t)\| + \chi(y) \tag{4.1}$$

Since $P(t)$ is bounded, by hypothesis, it follows that $\chi(x) \leq \chi(y)$. On the other hand, $y(t) = P(t)^{-1}x(t)$ and by hypothesis $P(t)^{-1}$ is also bounded, whence $\chi(y) \leq \chi(x)$ and finally $\chi(x) = \chi(y)$. A similar argument shows that $\nu(\chi(x)) = \nu(\chi(y))$. In other words, the numbers χ_j, ν_{jw}, $\mu(\chi_j)$, $\mu(\nu_{jw})$ are the same for $A(t)$ and $B(t)$.

The development of the idea that real Jordan matrices are canonical forms for a subset of M_{n+} will proceed in stages. Theorem **4.2** does not quite solve the problem although the link between Jordan matrices and kinematic similarity has been forged as a result of the invariants chosen. First of all we consider the simplest case and state the following well known result:

Theorem 4.3: *Static similarity - constant matrices*

Each constant matrix $B \in \overline{M}_n$ is statically, and thereby kinematically similar to a Jordan matrix $B_J \in \overline{M}_n$.

Proof: This is a well known result. For a proof see Gantmacher (1959). The assertion is that there exists a constant Liapunov transformation $P \in \overline{M}_n$ such that $B \in B_J$. That is

$$B_J = -P^{-1}[-BP]$$

$$= P^{-1}BP \tag{4.2}$$

It is interesting to note that static similarity implies kinematic similarity, although the converse is not true. The question here is one of uniqueness, as may be verified by means of simple examples. To clarify the situation consider the following:

Theorem 4.4: *Kinematic similarity - real Jordan matrices (Markus, 1955)*

Let A_J and B_J be real, constant, Jordan matrices. Arrange the order of the blocks A_j down the principal diagonal of A_J (say, first by increasing $|\lambda|$, second by increasing $\operatorname{Arg}\lambda$ and third by $\dim A_{jp}$; see Chapter **3**). If $A_J \sim B_J$ then $A_J = B_J$.

Proof: It is sufficient to verify that the invariants χ_j, ν_{jw}, $\mu(\chi_j)$, $\mu(\nu_{jw})$ determine the total structure of A_J. For details see section **3.4**.

Thus we are led to a stricter version of Theorem **4.3**, namely:

Theorem 4.5: *Kinematic similarity - constant matrices*

Subject to the ordering assumed above, each constant matrix $B \in \overline{M}_n$ is kinematically similar to a *real* Jordan matrix $B_J \in \overline{M}_n$.

Proof: By virtue of Theorem **4.3**, the constant matrix B is statically and therefore kinematically similar to a possibly *complex* Jordan matrix $\hat{B}_J \in \overline{M}_n$. Since Theorem **4.4** guarantees uniqueness, subject to the ordering of submatrix blocks, it is sufficient to prove that $\hat{B}_J \in B_J$. In fact, it is only necessary to consider elementary divisor blocks of the respective matrices. The assertion is that

$$(\hat{B}_J)_{jp} = \begin{pmatrix} \hat{b}_j & 1 & & 0 \\ & \hat{b}_j & \cdot & \\ & & \cdot & \cdot \\ & & & 1 \\ 0 & & & \hat{b}_j \end{pmatrix} \sim \begin{pmatrix} b_j & 1 & & 0 \\ & b_j & \cdot & \\ & & \cdot & \cdot \\ & & & 1 \\ 0 & & & b_j \end{pmatrix}$$

$$= (B_J)_{jp} \tag{4.3}$$

where $\hat{b}_j = b_j + i\xi_j$. Define the corresponding block of the assumed Liapunov transformation by

$$(\Pi(t))_{jp} = \exp(i\xi_j t) \begin{pmatrix} 1 & & & 0 \\ & 1 & & \\ & & \cdot & \\ 0 & & & 1 \end{pmatrix}$$

Direct calculation shows that

$$-(\Pi(t)^{-1})_{jp}[(\dot{\Pi}(t))_{jp} - (\hat{B}_J)_{jp}(\Pi(t))_{jp}] = (B_J)_{jp} \tag{4.4}$$

which completes the proof.

From this theorem we conclude that under kinematic similarity the unique canonical form B_J for B is obtained by taking a complex Jordan matrix $\hat{B}_J \sim B$ and then deleting the imaginary parts of the characteristic values of $\hat{B}_J$.

Theorem 4.6: *Complete set of invariants under kinematic similarity (Markus, 1955)*

For the subclass of matrices in M_{n+} which are kinematically similar to constant matrices, the invariants consisting of characteristic exponents χ_j, types ν_{jw}, and their multiplicities $\mu(\chi_j)$ and $\mu(\nu_{jw})$ form a *complete set of invariants*. Furthermore, if $A(t) \sim B$ for $A(t) \in M_{n+}$ and $B \in \overline{M}_n$, then there exists a unique real constant Jordan matrix $B_J \sim B$ which displays the invariants of A(t).

Proof: follows from Theorems 4.2 and 4.5.

4.4 Necessary and Sufficient Conditions for Kinematic Similarity

To fully exploit the concept of kinematic similarity it is desirable to have a set of necessary and sufficient conditions for kinematic similarity. Unfortunately the general problem here seems quite difficult. However, there are special situations for which it is possible to give conditions sufficient to ensure that two given matrices are kinematically similar. We treat the simplest case first.

Theorem 4.7: *Erugin's Theorem (Erugin, 1946)*

Let $A, B \in \overline{M}_n$. Then $A \sim B$ if and only if A_J and B_J have the same distribution of 1's on their superdiagonals, and for corresponding characteristic values a_j of A and b_j of B we have $\mathrm{Re}(a_j) = \mathrm{Re}(b_j)$, $j = 1,2,\ldots,n$.

Proof: follows from Theorems 4.4 and 4.5. Note that the requirement on the characteristic values simply means that the two matrices have the same characteristic exponents.

A more interesting case is that of matrices with time dependent elements kinematically similar to constant matrices. It is clear from the definition of kinematic similarity that the matrix

$A(t) \in M_{n+}$ is kinematically similar to a constant matrix $B \in \overline{M}_n$ if and only if there exists a Liapunov transformation $P(t)$ which satisfies

$$\dot{P}(t) = A(t)P(t) - P(t)B \tag{4.5}$$

for all $t \in R_+$. It is easily shown that if $P(t)$ is a solution of (4.5), then

$$X(t) = P(t)\exp[Bt]$$

is a fundamental matrix of

$$\dot{X}(t) = A(t)X(t) \tag{4.6}$$

and if $X(t)$ is a fundamental matrix of (4.6), then

$$P(t) = X(t)\exp[-Bt]$$

is a solution of (4.5). For $X(t)$ and $P(t)$ related in this way, it follows that

$$\det P(t) = \det X(t) \det \exp[-Bt] \tag{4.7}$$

and from Lemma **3.2** (Abel-Jacobi-Liouville)

$$\det P(t) = \det X(0)\exp\left[\int_0^t \operatorname{tr}(A(s)-B)ds\right] \tag{4.8}$$

If $P(t)$ is a solution of (4.5) and $X(t) = P(t)\exp[Bt]$ then $P(t)^{-1}$ exists if and only if the columns of $X(t)$ are linearly independent over the field of complex numbers since this is true if and only if $\det X(0) \neq 0$. Also we observe that if $P(t)^{-1}$ exists, then the columns of $P(t)$ are certainly linearly independent over the complex numbers. Moreover, if $P(t) \in M_{n+}$ is a solution of (4.5) such that $P(t)^{-1}$ exists, then $P(t)^{-1} \in M_{n+}$ if and only if

$$\operatorname{Re}\left(\int_0^t \operatorname{tr}(A(s)-B)ds\right) \tag{4.9}$$

is bounded for all $t \in R_+$. Clearly

$$\det P(t) \det P(t)^{-1} = 1.$$

If $P(t)^{-1}$ is bounded then $\det P(t)$ is bounded away from zero so that this, together with the boundedness of $P(t)$, implies through (4.8) that (4.9) holds, i.e.

$$\mathrm{Re}\left(\int_0^t \mathrm{tr}(A(s)-B)\,ds\right)$$

is bounded for all $t \in R_+$. Conversely, if (4.9) holds, then $\det P(t)$ is bounded away from zero and this, together with the boundedness of $P(t)$, implies that $P(t)^{-1} \in M_{n+}$. Note that according to Theorem **4.5** we may write

$$\mathrm{Re}\left(\int_0^t \mathrm{tr}(A(s)-B_J)\,ds\right) \tag{4.10}$$

in place of (4.9), where $B_J \in \overline{M}_n$ is a real Jordan matrix. By combining this argument and Theorem **4.6**, Langenhop *(op.cit.)* proves the following:

Theorem 4.8: *Necessary and sufficient conditions for $A(t) \sim B$ (Langenhop, 1960)*

Let $A(t) \in M_{n+}$ and $B \in \overline{M}_n$. Then for $A(t) \sim B$ it is necessary and sufficient that there exist real numbers χ_j which are characteristic exponents of $A(t)$ and which have multiplicities $\mu(\chi_j)$ with

$$\sum_{j=1}^{k} \mu(\chi_j) = n,$$

such that

$$\mathrm{Re}\left[\int_0^t \left(\mathrm{tr}\,A(s) - \sum_{j=1}^{k} \mu(\chi_j)\chi_j\right)ds\right] \tag{4.11}$$

is bounded for all $t \in R_+$.

Proof: for details see Langenhop *(op.cit.)*.

We call attention to the fact that Langenhop proves the above theorem for the whole real line. We have set it in R_+ to follow the already established trend in this chapter, knowing that it is generally possible to restate the results so far for R.

Theorem **4.8** contains the most general result in this section. Specific results of wide interest and application will be considered in the next section. The overall problem is essentially one of estimating the characteristic exponents and their multiplicities.

4.5 Estimates for Characteristic Exponents

We consider the simplest case first of all, that is let the kinematic similarity be defined for matrices $A(t) \in M_{1+}$ and $B \in \overline{M}_1$. An obvious possibility is that

$$\frac{1}{t}\int_0^t A(s)ds = B + \eta(t) \tag{4.12}$$

where $|\eta(t)| \to 0$ as $t \to \infty$. Then

$$B = \lim_{t\to\infty} \frac{1}{t}\int_0^t A(s)ds$$

In this case the Liapunov transformation is a bounded solution of (4.5) and is written as

$$P(t) = \exp\left[\int_0^t (A(s) - \lim_{s\to\infty} \frac{1}{s}\int_0^s A(\xi)d\xi)ds\right]$$

Clearly this result is applicable to those cases which can be treated as if $n = 1$, e.g. diagonal matrices, provided of course that the limit

$$\lim_{t\to\infty} \frac{1}{t}\int_0^t A(s)ds$$

exists. In the general case the relation $A(t) \sim B$ cannot be found by analogy with the scalar case. Recalling that the relation is defined by

$$B = -P(t)^{-1}[\dot{P}(t) - A(t)P(t)]$$

we obtain the linear differential equation

$$\dot{P}(t) = A(t)P(t) - P(t)B$$

whose solution is a Liapunov transformation given by

$$P(t) = X(t)\exp[-Bt]$$

where $X(t)$ satisfies the linear homogeneous equation

$$\dot{X}(t) = A(t)X(t).$$

Assume for the sake of argument that the matrix $B \in \overline{M}_n$ is defined by

$$B = \lim_{t\to+\infty} \frac{1}{t}\int_0^t A(s)ds \tag{4.13}$$

provided of course that the limit exists. Then, reasoning as for the scalar case, we are tempted to suggest that

$$P(t) = \exp\left[\int_0^t [A(s) - \lim_{s\to+\infty} \frac{1}{s}\int_0^s A(\xi)d\xi]ds\right] \tag{4.14}$$

This is generally incorrect for two reasons:

(i) it is only possible to express $X(t)$ as

$$X(t) = \exp\left[\int_0^t A(s)ds\right]$$

in certain cases, which will be specified later.

(ii) Should it happen that

$$X(t) = \exp\left[\int_0^t A(s)ds\right]$$

then

$$\begin{aligned} P(t) &= X(t)\exp[-Bt] \\ &= \exp\left[\int_0^t A(s)ds\right]\exp[-Bt] \end{aligned}$$

is not always equal to (4.14), since

$$\exp[A]\exp[B] \neq \exp[A+B] \quad \text{unless} \quad AB = BA.$$

As may easily be verified, if

$$A(t)\int_0^t A(s)ds = \int_0^t A(s)ds\, A(t)$$

then neither (i) nor (ii) hold. Furthermore, if

$$B = \lim_{t\to\infty} \frac{1}{t}\int_0^t A(s)ds \tag{4.13}$$

exists, then P(t) is given by (4.14).

In cases where the above limit does not exist, Vul'pe (1972) has succeeded in extending the above result by introducing a generalised limit $\bar{B} \in \bar{M}_n$ given by

$$\bar{B} = \lim_{t\to\infty} \frac{1}{t}\int_0^t \left[\frac{1}{s}\int_0^s A(\xi)d\xi\right]ds \tag{4.15}$$

which has the property that if B exists, then $\bar{B}$ exists and $B = \bar{B}$; however the converse does not necessarily hold.

We summarize the above discussion in the following:

Theorem 4.9

Let $A(t) \in M_{n+}$ commute with $\int_0^t A(s)ds$ for each $t \in R_+$. Assume that A(t) can be decomposed into $A(t) = A_p(t)+A_o$, where $A_o \in \bar{M}_n$ and commutes with both A(t) and $\int_0^t A(s)ds$. Suppose that $\int_0^t A_p(s)ds \in M_{n+}$. Then $A(t) \sim A_o$ and

$$A_o = \lim_{t\to\infty} \frac{1}{t}\int_0^t A(s)ds.$$

Proof: To begin with certain commutation results must be established. First of all

$$A_p(t)\int_0^t A_p(s)ds = (A(t)-A_o)\left(\int_0^t A(s)ds - \int_0^t A_o ds\right)$$

$$= \left(\int_0^t A(s)ds - \int_0^t A_o ds\right)(A(t)-A_o)$$

$$= \int_0^t A_p(s)ds\ A_p(t) \qquad (4.16)$$

Similarly

$$A_o \int_0^t A_p(s)ds = A_o\left(\int_0^t A(s)ds - \int_0^t A_o ds\right)$$

$$= \left(\int_0^t A(s)ds - \int_0^t A_o ds\right) A_o$$

$$= \int_0^t A_p(s)ds\ A_o \qquad (4.17)$$

Let

$$P(t) = \exp\left(\int_0^t A_p(s)ds\right).$$

It is easily shown that

$$-P(t)^{-1}(\dot{P}(t)-A(t)P(t)) = A_o.$$

Since $\int_0^t A_p(s)ds \in M_{n+}$, both $P(t)$ and $P(t)^{-1} \in M_{n+}$, and clearly $P(t)$ possesses a continuous first derivative. Therefore $A(t) \sim A_o$ as required.

To clarify the conditions under which a matrix commutes with its integral we introduce

$$B(t) = \int_0^t A(s)ds \qquad (4.18)$$

where $A(t), B(t) \in M_{n+}$. Assuming that

(i) there exists a nonsingular differentiable matrix $P(t)$ such that

$$B(t) = P(t)^{-1}B_J(t)P(t) \tag{4.19}$$

where $B_J(t) \in M_{n+}$ is in Jordan canonical form,

(ii) the distribution of 1's on the superdiagonal of $B_J(t)$ does not change for all $t \in R_+$,

(iii) no difference between different characteristic values of $B_J(t)$ vanishes in a subinterval of R_+ unless it vanishes identically,

(iv) if the difference between different characteristic values of $B_J(t)$ vanishes identically then the superdiagonals in either elementary divisor block become 0.

Epstein has proved the following:

Theorem 4.10: *(Epstein, 1963)*

The matrix $B(t) \in M_{n+}$ which satisfies

$$\dot{B}(t)B(t) - B(t)\dot{B}(t) = 0,$$

that is

$$A(t)\int_0^t A(s)ds - \int_0^t A(s)dsA(t) = 0 \tag{4.20}$$

and having a Jordan canonical form $B_J(t) \in M_{n+}$ whose structure is constant for all $t \in R_+$, is obtained

(i) by finding all matrices $X(t)$ satisfying

$$B_J(t)[X(t)B_J(t)-B_J(t)X(t)]-[X(t)B_J(t)-B_J(t)X(t)]B_J(t) = 0 \tag{4.21}$$

(ii) determining the nonsingular solutions $P(t)$ of the matrix differential equation

$$\dot{P}(t) = X(t)P(t) \tag{4.22}$$

(iii) forming

$$B(t) = P(t)^{-1}B_J(t)P(t) \tag{4.23}$$

The matrices $X(t)$ form a linear space under addition which depends only on the structure of the elementary divisor blocks and the set of subscript pairs for which the difference between different characteristic values vanishes.

Proof: Observe that, trivially

$$\dot{B}_J(t)B_J(t) = B_J(t)\dot{B}_J(t) \tag{4.24}$$

Differentiating (4.23) yields

$$\dot{B}(t) = \dot{P}(t)^{-1}B_J(t)P(t) + P(t)^{-1}\dot{B}_J(t)P(t) + P(t)^{-1}B_J(t)\dot{P}(t) \tag{4.25}$$

whence

$$\dot{B}(t)B(t)-B(t)\dot{B}(t) = P(t)^{-1}[-X(t)B_J(t)+\dot{B}_J(t)+B_J(t)X(t)]B_J(t)P(t)$$

$$- P(t)^{-1}B_J(t)[-X(t)B_J(t)+\dot{B}_J(t)+B_J(t)X(t)]P(t) \tag{4.26}$$

If

$$\dot{B}(t)B(t) - B(t)\dot{B}(t) = 0,$$

then

$$[-X(t)B_J(t)+\dot{B}_J(t)+B_J(t)X(t)]B_J(t)$$

$$- B_J(t)[-X(t)B_J(t)+\dot{B}_J(t)+B_J(t)X(t)] = 0 \tag{4.27}$$

Taking into account (4.24), (4.26) becomes

$$B_J(t)[X(t)B_J(t)-B_J(t)X(t)]-[X(t)B_J(t)-B_J(t)X(t)]B_J(t) = 0$$

which is precisely (4.21).

An immediate consequence of Theorem **4.10** is the following:

Theorem 4.11: *(Epstein, 1963)*

Let $A(t) \in M_{n+}$ commute with its integral, that is

$$A(t)\int_0^t A(s)ds = \int_0^t A(s)ds\, A(t).$$

Then

$$A(t) \sim [X(t)+\dot{B}_J(t)+B_J(t)X(t)-X(t)B_J(t)] \tag{4.28}$$

where $X(t)$, $B_J(t)$ are defined as in Theorem **4.10**. The Liapunov transformation is of course $P(t)$.

Proof: follows directly from (4.25).

For the particular class of matrices which are *normal*, that is statically similar, by a unitary transformation, to a complex diagonal matrix, it is possible to derive a relation between the characteristic exponents of $A(t) \in M_{n+}$ and the averages of the characteristic values of $A(t)$ for each fixed value of t.

Theorem 4.12: *(Markus, 1955)*

Let $A(t) \in M_{n+}$ be normal for each fixed t on R_+. Let $\Lambda(t)$ be the maximum of the real parts of the characteristic values of $A(t)$ and $\lambda(t)$ be the minimum of these real parts. Then,

$$\overline{\lim_{t\to\infty}} \frac{1}{t} \int_0^t \lambda(s)ds \;\le\; \chi_j \;\le\; \overline{\lim_{t\to\infty}} \frac{1}{t} \int_0^t \Lambda(s)ds \tag{4.29}$$

where χ_j are the characteristic exponents of $A(t)$.

Proof: It is known (Hamburger & Grimshaw, 1951) that $\Lambda(t)$ and $\lambda(t)$ are real, continuous and bounded functions on R_+. For any solution $x(t)$ of $\dot{x}(t) = A(t)x(t)$ define

$$r(t)^2 = x(t)^* x(t) \tag{4.30}$$

Differentiating, we have

$$\begin{aligned} 2r(t)\dot{r}(t) &= x(t)^*\dot{x}(t) + \dot{x}(t)^* x(t) \\ &= x(t)^* A(t)x(t) + x(t)^* A(t)^* x(t) \end{aligned} \tag{4.31}$$

whence

$$\dot{r}(t) = (x(t)^* A(t)x(t) + x(t)^* A(t)^* x(t))r(t)(2x(t)^* x(t))^{-1}$$

Since $A(t)$ is normal for each $t \in R_+$

$$\begin{aligned} \lambda(t) &\le (x(t)^* A(t)x(t)+x(t)^* A(t)^* x(t))(2x(t)^* x(t))^{-1} \\ &\le \Lambda(t) \end{aligned} \tag{4.32}$$

and

$$\lambda(t) \le \dot{r}(t)r(t)^{-1} \le \Lambda(t) \tag{4.33}$$

Integrating, we obtain

$$\int_0^t \lambda(s)ds \le \log r(t) - \log r(0) \le \int_0^t \Lambda(s)ds \tag{4.34}$$

Now

$$\chi = \overline{\lim_{t\to\infty}} \frac{1}{t} \log\|x(t)\| = \overline{\lim_{t\to\infty}} \frac{1}{t} \log r(t) \tag{4.35}$$

Therefore

$$\overline{\lim_{t\to\infty}} \frac{1}{t}\int_0^t \lambda(s)ds \le \overline{\lim_{t\to\infty}} \frac{1}{t}\log r(t) \le \overline{\lim_{t\to\infty}} \frac{1}{t}\int_0^t \Lambda(s)ds$$

or

$$\overline{\lim_{t\to\infty}} \frac{1}{t}\int_0^t \lambda(s)ds \le \chi \le \overline{\lim_{t\to\infty}} \frac{1}{t}\int_0^t \Lambda(s)ds$$

which proves the theorem.

Observe that if

$$\overline{\lim_{t\to\infty}} \frac{1}{t}\int_0^t \Lambda(s)ds < 0,$$

then every solution of $\dot{x} = A(t)x$ under the assumptions of the theorem, approaches the origin as $t \to \infty$.

Theorems 4.9 and 4.12 are quite general and we take advantage of this fact to derive some interesting results for almost periodic matrices

Theorem 4.13: *(Markus, 1955)*

If $A(t) \in AP_{n+}$ (i.e. $A(t) \in M_{n+}$, $A(t)$ is almost periodic) and $A(t)$, $\int_0^t A(s)ds$ and A_0 all commute on R_+ with $A_0 = M_t\{A\}$. If

$$\int_0^t (A(s)-A_0)ds \in M_{n+}$$

then $A(t) \sim A_0$. Moreover, the Liapunov transformation $P(t)$ is almost periodic.

Proof: follows from Theorem 4.9. (For the definition and properties of the mean value operator M_t see Chapter 2.)

In the same spirit we have the following:

Theorem 4.14: *(Markus, 1955)*

If $A(t) \in AP_{n+}$ and $A(t)$, $\int_0^t A(s)ds$ commute on R_+ and $A(t)$ has an absolutely convergent Fourier series with exponents bounded away from zero, then $A(t) \sim A_o = 0$. Moreover, the Liapunov transformation is almost periodic.

Proof: By hypothesis $\int_0^t A(s)ds$ is bounded and thus almost periodic. Since

$$\lim_{t\to\infty} \frac{1}{t} \int_0^t A(s)ds = 0$$

the result follows immediately.

To establish bounds on the characteristic exponents of an almost periodic normal matrix, we use Theorem 4.12, which can be restated as:

Theorem 4.15: *(Markus, 1955)*

Let $A(t) \varepsilon M_{n+}$ be normal for each fixed $t \varepsilon R_+$ and let $A(t)$ be almost periodic or have period $T > 0$. Then $\Lambda(t)$ and $\lambda(t)$ are also almost periodic, or have period T respectively, and

$$\lim_{t\to\infty} \frac{1}{t} \int_0^t \lambda(s)ds \le \chi_j \le \lim_{t\to\infty} \frac{1}{t} \int_0^t \Lambda(s)ds \qquad (4.36)$$

Proof: Let $A(t)$ have period $T > 0$. Then clearly both $\Lambda(t)$ and $\lambda(t)$ have period T.

Suppose that $A(t)$ is almost periodic and let τ be an η-almost period of $A(t)$, i.e. $\|A(t+\tau)-A(t)\| < \eta$ for all $t \varepsilon R$. Since $\|A(t)\|$ is bounded, we can choose η so small that $|\Lambda(t+\tau) - \Lambda(t)| < \delta$ and $|\lambda(t+\tau)-\lambda(t)| < \delta$ for a preassigned $\delta > 0$ and all $t \varepsilon R$. Thus both $\Lambda(t)$ and $\lambda(t)$ are almost periodic. Therefore the limits proposed in (4.36) do exist and the required result follows directly from Theorem 4.12.

The problem of estimating the characteristic exponents of a periodic matrix has attracted considerable attention in the literature (Proctor, 1969; Sansone & Conti, 1964; Bellman, 1953; Gantmacher, 1959). Floquet's Theorem (Theorem 3.8) gives the most obvious result, first stated by Liapunov.

Theorem 4.16: *(Liapunov, 1893)*

Let $A(t) \varepsilon M_{n+}$ be periodic with period $T > 0$. Then $A(t) \sim B$ with $B \varepsilon \overline{M}_n$ given by

$$B = T^{-1} \log X(T) \tag{4.37}$$

where $X(t)$ is a solution of $\dot{X}(t) = A(t)X(t)$. Moreover, the Liapunov transformation $P(t)$ is periodic with period T.

Proof: follows from Theorem **3.8**.

It is unfortunate that the matrix differential equation $\dot{X}(t) = A(t)X(t)$ must be solved before B can be constructed according to (4.37). In some circumstances it may be possible to construct B using a successive approximation technique. The idea is to solve the problem $A(t)\delta \sim B$ where δ is a small parameter and then see if the solution is valid for $\delta = 1$. According to Theorems **4.16** and 3.8, a periodic Liapunov transformation $P(t)$ is sought in the form

$$P(t) = X(t)\exp[-Bt] \tag{4.38}$$

Assume that $X(t)$ can be expressed as a power series in the parameter δ

$$X(t) = \sum_{k=0}^{\infty} X_k(t)\delta^k, \qquad X_o = I \tag{4.39}$$

This converges for all finite values of δ and $t \in R$. At $t = T$,

$$X(T) = \sum_{k=0}^{\infty} X_k(T)\delta^k \tag{4.40}$$

by definition. Therefore

$$B = T^{-1}\log X(T)$$

can be written as

$$B = \sum_{k=0}^{\infty} B_k\delta^k \tag{4.41}$$

The convergence of this series is established by considering a corresponding majorant series. From the relation

$$P(t) = \sum_{k=0}^{\infty} X_k(t)\delta^k \exp(-\sum_{k=0}^{\infty} B_k\delta^k t), \qquad X_o = I \tag{4.42}$$

it is clear that $P(t)$ can be written as

$$P(t) = I + \sum_{k=1}^{\infty} P_k(t)\delta^k \tag{4.43}$$

which has the same radius of convergence as (4.41). Substituting the expressions for P(t) and B into the differential equation

$$\dot{P}(t) = A(t)P(t)\delta - P(t)B \tag{4.44}$$

and equating like powers of δ yields

$$\dot{P}_k(t) = A(t)P_{k-1}(t) - \sum_{m=1}^{k-1} P_{k-m}(t)B_m - B_k \tag{4.45}$$

$$\dot{P}_1(t) = A(t) - B_1 \tag{4.46}$$

Since the matrices $A(t), P_1(t) \in M_{n+}$ are real and periodic by hypothesis, it follows from (4.46) that

$$P_1(t) = \int_0^t [A(s)-B_1]ds$$

and

$$B_1 = \frac{1}{T}\int_0^T A(s)ds$$

Therefore

$$P_1(t) = \int_0^t A(s)ds - \frac{t}{T}\int_0^T A(s)ds \tag{4.47}$$

Similarly an expression for $P_2(t)$ can be obtained by the same procedure and in general

$$B_k = \frac{1}{T}\int_0^T [A(s)P_{k-1}(s) - \sum_{m=1}^{k-1} P_{k-m}(s)B_m]ds \tag{4.48}$$

and

$$P_k(t) = \int_0^t [A(s)P_{k-1}(s) - \sum_{m=1}^{k-1} P_{k-m}(s)B_m]ds - B_k t \tag{4.49}$$

Hence the series for P(t) and B may be constructed and will converge for sufficiently small values of δ. Note that the series will not converge for any values of δ for which the matrix X(T) has characteristic values with negative real parts, since we have assumed throughout that the Liapunov transformation has period T.

In the remainder of this chapter we shall consider linear time-varying systems with bounded and continuous coefficients,

$$\dot{x} = A(t)x, \qquad A(t) \in M_n \tag{4.50}$$

which are commutative, that is

$$A(t)\int_{t_o}^{t} A(s)ds = \int_{t_o}^{t} A(s)dsA(t)$$

for all t and establish conditions for the evaluation of the state transition matrix $\Phi(t,t_o)$ and hence the stability of (4.50).

The coefficient matrix $A(t) \in M_n$ can always be decomposed into the non-unique representation (Wu, 1980)

$$A(t) = \sum_{j=1}^{r} f_j(t)F_i \tag{4.51}$$

where $\{f_i(t)\}_1^r$ are linearly independent scalar sets of functions of $t \in R$ which are extracted from A(t) and $F_i \in \bar{M}_n$. In this case the commutative property of A(t) yields

$$A(t)\left(\int_{t_o}^{t} A(s)ds\right) = \left(\sum_{j=1}^{r} f_j(t)F_j\right)\left(\int_{t_o}^{t} \sum_{j=1}^{r} f_j(s)F_j ds\right)$$

$$= \left(\sum_{j=1}^{r} f_j(t)F_j\right)\left(\sum_{j=1}^{r} h_j(t,t_o)F_j\right)$$

$$= \int_{t_o}^{t} A(s)dsA(t)$$

$$= \left(\sum_{j=1}^{r} h_j(t,t_o)F_j\right)\left(\sum_{j=1}^{r} f_j(t)F_j\right)$$

which can only hold iff the $\{F_j\}_1^r$ are mutually commutative, that is,

$$F_iF_j = F_jF_i \qquad \text{for all} \quad i,j = 1,2,\dots r,$$

where

$$h_j(t,t_o) = \int_{t_o}^{t} f_j(s)ds.$$

Also since the linear system (4.50) is commutative

$$\Phi(t,t_o) = \exp\left[\int_{t_o}^{t} A(s)ds\right]$$

$$= \exp\left[\sum_{j=1}^{r} h_j(t,t_o)F_j\right]$$

$$= \prod_{j=1}^{r} \exp[h_j(t,t_o)F_j] \tag{4.52}$$

and from $\{F_j\} \varepsilon \bar{M}_n$ the state transition matrix of (4.50) can be computed through (4.52) as though the system was time-invariant, since $h_j(t,t_o)$ are scalars. We note in passing that if each F_j has n-distinct characteristic values $\lambda_k[F_j]$, $k = 1,\dots,n$; Theorem **4.3** gives

$$P^{-1}F_iP = \text{Diag}\{\lambda_k[F_i]\}$$

for $P \varepsilon \bar{M}_n$ a nonsingular Liapunov transformation (or similarity transformation in this case). Hence

$$P^{-1}A(t)P = P^{-1}[\sum_{j=1}^{r} f_j(t)F_j]P$$

$$= \sum_{j=1}^{r} f_j(t)P^{-1}F_jP = \sum_{j=1}^{r} f_j(t)\text{Diag}\{\lambda_k[F_j]\} \tag{4.53}$$

The transformation of A(t) has diagonalised both sides of (4.53) and since A(t) and $P^{-1}A(t)P$ must have the same characteristic values $\{\lambda_j(A(t))\}$, then (4.53) provides a simple relationship between the characteristic values of $A(t)$ and the constant matrices F_i and the associated functions $f_i(t)$, as,

$$\lambda_i(A(t)) = \sum_{j=1}^{r} f_j(t)\lambda_i[F_j], \qquad i = 1,2,\ldots n. \tag{4.54}$$

We can summarise the above in the following theorem:

Theorem 4.16

Consider the linear time varying system $\dot{x} = A(t)x$, if $A(t) \in M_n$ is commutative then

(i) $A(t)$ can be expressed as $\sum_{j=1}^{r} f_j(t)F_j$ where $\{F_j\}_1^r$ are mutually commutative,

(ii) the system state transition matrix can be written as

$$\Phi(t,t_o) = \prod_{j=1}^{r} \exp[h_j(t,t_o)F_j]$$

where $h_j(t,t_o) = \int_{t_o}^{t} f_j(s)ds$. The necessary and sufficient conditions for stability can be determined directly from this expression for $\Phi(t,t_o)$.

(iii) If the characteristic values of $A(t)$ are distinct then they are given by

$$\lambda_i(A(t)) = \sum_{j=1}^{r} f_j(t)\lambda_i[F_j], \qquad i = 1,2,\ldots n.$$

(iv) The system $\dot{x} = A(t)x$, is (a) *stable* if $|h_i(t,t_o)| < \infty$ for all (t,t_o) and $i = 1,2,\ldots,r$; (b) *asymptotically stable* if $\lim_{t\to\infty} h_i(t,t_o) \to \infty$ with $h_i(t,t_o) > 0$ for all (t,t_o) and $i = 1,2,\ldots,r$; and $\mathrm{Re}\{\lambda_k[F_j]\} < 0$ for all $j = 1,2,\ldots,r$; $k = 1,2,\ldots,n$.

Example 4.1

Consider the linear periodic system $\dot{x} = A(t)x$ with

$$A(t) = \begin{bmatrix} \alpha\cos\omega t & \beta\sin\omega t \\ -\delta\sin\omega t & \alpha\cos\omega t \end{bmatrix} \tag{4.55}$$

for $\delta,\beta > 0$, α,δ,β finite and $\omega \neq 0$. Clearly $A(t)$ can be decomposed into the form (4.51) as

$$A(t) = \alpha\cos\omega t \begin{bmatrix} 1 & 0 \\ 0 & 1 \end{bmatrix} + \sin\omega t \begin{bmatrix} 0 & \beta \\ -\delta & 0 \end{bmatrix}$$

$$= f_1(t)F_1 + f_2(t)F_2 \ .$$

The matrices $F_1, F_2 \in \bar{M}_n$ clearly commute and so by theorem **4.16** the periodic system above is commutative. Also the system (4.55) is stable since

$$|f_1(t)| = |\alpha\cos\omega t| < \infty$$

and

$$|f_2(t)| = |\sin\omega t| < \infty \qquad \text{for all} \quad (\omega,t).$$

From equation (4.52) the state transition matrix of system (4.55) is easily computed on noting that, for $t_o = 0$

$$h_1(t,0) = \int_o^t \alpha\cos\omega s ds = \alpha\omega^{-1}\sin\omega t,$$

$$h_2(t,0) = \int_o^t \sin\omega s ds = \omega^{-1}(1-\cos\omega t),$$

as

$$\Phi(t,0) = \exp(\alpha\omega^{-1}\sin\omega t)\begin{bmatrix} \cos(\delta\beta\omega^{-1}(1-\cos\omega t)) & \delta^{-1}\sin(\delta\beta\omega^{-1}(1-\cos\omega t)) \\ -\beta^{-1}\sin(\delta\beta\omega^{-1}(1-\cos\omega t)) & \cos(\delta\beta\omega^{-1}(1-\cos\omega t)) \end{bmatrix}$$

The characteristic values of F_1 are (1,1) and the characteristic values of F_2 are $\pm j(\delta\beta)^{\frac{1}{2}}$, so from theorem **4.16** the characteristic values of $A(t)$ are given by

$$\lambda_1(A(t)) = (\alpha^2+\delta^2\beta^2)^{\frac{1}{2}}\exp(j\omega t)$$

$$\lambda_2(A(t)) = (\alpha^2+\delta^2\beta^2)^{\frac{1}{2}}\exp(-j\omega t)$$

which indicates that the characteristic values of A(t) are periodic and alternate between the left and right half s-plane; however the commutative periodic system $\dot{x} = A(t)x$ is always stable as long as $\omega \neq 0$, and so stability is independent of the characteristic values of $\{F_i\}$ whenever $\{f_i(t)\} \not\subset \bar{M}_1$. When $\omega = 0$ the periodic system (4.55) becomes time-invariant and *unstable*, which indicates that unstable linear time-invariant systems can be stabilised by sinusoidal modulation.

The results of theorem **4.16** can be related to theorem **4.9**, for the decomposition $A(t) = A_p(t) + A_o \varepsilon M_{n+}$, $A_o \varepsilon \bar{M}_n$ by expanding $A_p(t)$ as,

$$A_p(t) = \sum_{j=1}^{m} g_j(t)G_j$$

The Liapunov transformation in this case becomes

$$P(t) = \exp\left[\int_o^t A_p(s)ds\right] = \prod_{j=1}^{m} \exp[h_j(t)G_j]$$

where $h_j(t) = \int_o^t g_j(s)ds$ for $j = 1,2,\ldots,m$. The state transition matrix for the system $\dot{x} = (A_p(t)+A_o)x$ follows from the above and theorem **4.16** as

$$\Phi(t,0) = \left\{\prod_{j=1}^{m} \exp[h_j(t)G_j]\right\}\exp A_o t \tag{4.56}$$

which is readily computed once the decomposition of $A_p(t)$ is fixed. Finally from theorem **4.9** the system $\dot{x} = A(t)x$ is kinematically similar to the time invariant system $\dot{y} = A_o y$.

A difficulty with theorem **4.16** is the non-uniqueness of the decomposition (4.51); however if the coefficient matrix $A(t) \varepsilon M_n$

has n-distinct characteristic values $\lambda_i(t)$, $i = 1,2,\ldots,n$; Zadeh and Desoer (1963) have shown that A(t) has a unique decomposition for all $t \in R$

$$A(t) = \sum_{i=1}^{n} \lambda_i(t) R_i(t) \tag{4.57}$$

where the (n×n) matrix $R_i(t) \in M_n$ is given by

$$R_i(t) = \lim_{s\to\lambda_i(t)} \{(s-\lambda_i(t))(sI-A(t))^{-1}\}$$

$$= e_i(t)(e_j'(t))^T \tag{4.58}$$

for all t. $R_i(t)$ is called the *residue matrix* of $(sI-A(t))^{-1}$ associated with the characteristic value $\lambda_i(t)$ and the set of characteristic vectors $\{e_i(t)\}_1^n$ of A(t); $\{(e_i'(t))^T\}_1^n$ is the transpose of the reciprocal basis of $\{e_i(t)\}_1^n$. From (4.58) we see that for all $t \in R$

$$R_i(t)R_j(t) = R_j(t)R_i(t) = \delta_{ij}R_i(t) \tag{4.59}$$

and so the residue matrices are mutually commutative, and the state transition matrix of the time-varying system $\dot{x} = A(t)x$ with $A(t) \in M_n$ commutative with its integral, is given by

$$\Phi(t,t_o) = \prod_{j=1}^{n} \left[\exp R_i \int_{t_o}^{t} \lambda_j(s)ds\right] \tag{4.60}$$

if A(t) is decomposed in its spectral form (4.57). This result is only of use if $R_i \in \bar{M}_n$ for all $t \in R$ and $i = 1,2,\ldots,n$.

Example 4.2: (Marcus-Yambe problem (1960))

Consider the linear periodic control system

$$\dot{x} = A(t)x + u \tag{4.61}$$

where $u = (0,u_2(t))^T$ and

$$A(t) = \begin{bmatrix} -1 - \alpha\cos^2 t & 1 - \alpha\sin t\cos t \\ -1 - \alpha\sin t\cos t & -1 + \alpha\sin^2 t \end{bmatrix} \tag{4.62}$$

The characteristic values of A(t) are $\{(\alpha-2)\pm(\alpha^2-4)\}/2$ which are independent of t. If the system (4.61) were stationary we would predict stability for $\alpha < 2$, in fact it is known that (4.61) possesses *unbounded* solutions for $\alpha < 2$, and this apparent contradiction is resolved as follows:-

Since A(t) is periodic it is kinematically similar to a constant matrix a suitable Liapunov transform is given by,

$$P(t) = \begin{bmatrix} \cos t & \sin t \\ -\sin t & \cos t \end{bmatrix}$$

so that the above transformation, $x(t) = P(t)y(t)$, produces the time invariant system

$$\dot{y} = By + E(t)u \tag{4.63}$$

where

$$B = \begin{bmatrix} \alpha-1 & 0 \\ 0 & -1 \end{bmatrix}, \qquad E(t) = \begin{bmatrix} \cos t & -\sin t \\ \sin t & \cos t \end{bmatrix},$$

and the condition for stability is $\alpha < 1$. Suppose however that $\alpha = 2$ then the kinematically similar systems (4.61), (4.63) are unstable; the problem is now to select a feedback $u(t) = -C(t)x$ such that

$$\dot{x} = [A(t)-C(t)]x \tag{4.64}$$

is stable. Using the same Liapunov transformation, P(t), it will be assumed that the constant matrix F is kinematically similar to [A(t)-C(t)] and given by

$$F = \begin{bmatrix} -2 & 0 \\ 0 & -1 \end{bmatrix}$$

That is $F \sim A(t)-C(t)$ has characteristic values -2 and -1 and therefore stable. It follows that

$$\dot{y}(t) = Fy + P(t)^{-1}u(t) \tag{4.65}$$

with $x(t) = P(t)y(t)$. However, F is composed of two parts B and some matrix D such that

$$F = B - D = \begin{bmatrix} \alpha-1 & 0 \\ 0 & -1 \end{bmatrix}_{\alpha=2} - \begin{bmatrix} d_{11} & d_{12} \\ d_{21} & d_{22} \end{bmatrix}$$

so that $d_{22} = d_{12} = d_{21} = 0$ and $d_{11} = 3$, for the selected F. Thence

$$[B-D] \sim A(t) - C(t)$$

or alternatively

$$P(t)BP(t)^{-1} + \dot{P}(t)P(t)^{-1} - P(t)DP(t)^{-1} = A(t) - C(t).$$

But $B \sim A(t)$, therefore

$$C(t) = P(t)DP(t)^{-1}$$

$$= 3\begin{bmatrix} \cos^2 t & -\sin t\cos t \\ -\sin t\cos t & \sin^2 t \end{bmatrix}$$

which illustrates that the unstable periodic system (4.61) can be stabilised by a state feedback control law with time varying gain C(t).

It is possible to investigate the stability of the system $\dot{x} = A(t)$ via another Liapunov transformation $P(t) = \exp(P_1 t)$, $P_1 \varepsilon \bar{M}_n$ if $A(t), \dot{A}(t) \varepsilon M_n$. We know that $A(t) \sim B$ if $B = P(t)^{-1} (A(t)P(t)-\dot{P}(t))$, so using the above Liapunov transformation

$$P^{-1}(t)(A(t)P(t)-\dot{P}(t)) = \exp(-P_1 t_o)(A(t_o)-P_1)\exp(P_1 t_o)$$

which is independent of t for all t_o and can be set equal to B. The solution to the time invariant system $\dot{y} = By$ is

$$y(t) = \exp\{B(t-t_o)\}y(t_o)$$

therefore

$$x(t) = P(t)y(t)$$

$$= \exp(P_1 t)\exp(B(t-t_o))y(t_o)$$

$$= \exp(P_1 t)\exp(B(t-t_o))\exp(-P_1 t_o)x(t_o)$$

$$\equiv \Phi(t,t_o)x(t_o) \tag{4.66}$$

Example 4.3

Consider again the Marcus-Yambe problem of example **4.2**. A suitable Liapunov transformation matrix for the periodic matrix A(t) of (4.62) is

$$P(t) = \exp\begin{bmatrix} 0 & 1 \\ -1 & 0 \end{bmatrix} t$$

so that $B \sim A(t)$ is given by

$$B = \exp\left\{\begin{bmatrix} 0 & -1 \\ 1 & 0 \end{bmatrix} t_o\right\}\left\{\begin{bmatrix} -1+\alpha\cos^2 t_o & -\alpha\sin t_o\cos t_o \\ -\alpha\sin t_o\cos t_o & -1+\alpha\sin^2 t_o \end{bmatrix}\right\}$$

$$\times \exp\left\{\begin{bmatrix} 0 & 1 \\ -1 & 0 \end{bmatrix} t_o\right\}$$

$$= \begin{bmatrix} \alpha-1 & 0 \\ 0 & -1 \end{bmatrix}$$

for $t_o = 0$; the same value for B as in example **4.2**.

Finally from (4.66), the state transition matrix is given by

$$\Phi(t,0) = \begin{bmatrix} \exp[(\alpha-1)t]\cos t & \exp(-t)\sin t \\ -\exp[(\alpha-1)t]\sin t & \exp(-t)\cos t \end{bmatrix}$$

which illustrates that the periodic system (4.61) is stable for $\alpha < 1$.

References

Bellman, R. (1953). "Stability theory of differential equations", McGraw-Hill, New York

Erugin, N.P. (1946). *Trud.Mat.Inst.Steklov* **13**, 95 (in Russian)

Gantmacher, F.R. (1959). "The theory of matrices", Vol.I, II, Chelsea, New York

Hamburger, H.L. and Grimshaw, M.E. (1951). "Linear transformation in n-dimensional vector space", Cambridge University Press

Langenhop, C.E. (1960). *Trans.Amer.Math.Soc.* **97**, 317-326

Liapunov, A.M. (1893). *Comm.Soc.Math.Kharkov* (in Russian).(In tran lation: *Ann.Math.Studies* **17**, Princeton, (1949))

Lillo, J.C. (1961). *Proc.Amer.Math.Soc.* **12**, 400-407

Markus, L. (1955). *Math.Zeitschr* **62**, 310-319

Markus, L. and Yambe, H. (1960). *Osaka Math.J.* **12**, 305-312

Martin, J.F.P. (1966). *Proc.Amer.Math.Soc.* **17**, 636-648

Proctor, T.G. (1969). *Proc.Amer.Math.Soc.* **22**, 503-508

Sansone, G. and Conti, R. (1964). "Nonlinear differential equations", Pergamon Press, London

Vulpé, I.M. (1972). *Diff.Uravneniya* **8**, 2156

Wu, M.-Y. (1980). *Int.J.Control* **31**, 11-20

Zadeh, L.A. and Desoer, C.A. (1963). "Linear systems theory", McGraw-Hill, New York

Chapter 5

STABILITY THEORY FOR NON-STATIONARY SYSTEMS

5.1 Local Equilibrium Stability Conditions

The stability of dynamical systems with respect to initial conditions and disturbances is the single most significant criterionin system design. The equilibrium states of a system (given by those points in the state space for which the time derivative of the state vector is zero) is stable if for small initial disturbances from this point the system remains within the vacinity of the equilibrium state. Should the system response eventually converge to this point then the stability property is also *asymptotic* within this domain of state space. Clearly these concepts are local and the *domain of attraction* or convergence about the equilibrium state is crucial in control system design. Hopefully this domain of attraction includes the whole state space, we then have *asymptotic stability in the large*, whereby after any initial disturbance and initial state the system response will eventually converge to the equilibrium state.

A very large number of definitions of stability exist, (LaSalle and Lefeschetz, 1961; Hahn, 1963, 1967; Willems, 1970; Yoshizawa, 1975; and Venkatesh, 1977) but only those which are of practical use and are relevant to systems described by homogeneous nonstationary differential equations will be discussed in this chapter. The various definitions of stability can be broadly classified as those which deal with the equilibrium of the null solution of free or unforced systems and those which consider the dynamic response of systems subject to various classes of forcing functions or inputs (these usually lead to input/output stability criteria - see Desoer and Vidysagar, 1975).

We will consider the general vector set of homogeneous differential equations,

$$\dot{x} = f(t,x), \quad x(t_o) = x_o \tag{5.1}$$

where x(t) is a n-dimensional continuously differentiable state vector with $x:R_+ = [o,\infty) \to E^n$, $f:R_+ \times B = D \to E^n$ with $f(t,o) = 0$ for all $t \in R_+$ and f(t,x) satisfies a Lipschitz condition in x and $B = \{x : x \in E^n, \ \|x\| < \alpha, \ \alpha > o\}$. In the following, the equilibrium state x_e of (5.1) can always be set equal to zero by a linear state transformation, so that the equilibrium state x_e and the null solution to (5.1) are considered throughout as equivalent.

Definition 5.1: *Stability of the equilibrium state (Willems, 1970)*

The zero solution $x(t;x_o,t_o) = 0$ for all $t \geq t_o$ of the differential equation $\dot{x} = f(t,x)$ is said to be *Liapunov stable* if and only if for each $t_o > 0$ and each $\xi > 0$ there is a $\delta(\xi,t_o) > 0$ such that

$$\|x_o\| < \delta(\xi,t_o)$$

implies that

$$\|x(t;x_o,t_o)\| < \xi \qquad \text{for all} \quad t \geq t_o.$$

The geometric interpretation of definition 5.1 is that by constraining the initial state to a sufficiently small sphere centred at the origin the resulting state trajectory remains for all $t > t_o$ in a prescribed sphere of radius ξ.

If the describing differential equations (5.1) are linear:

$$\dot{x} = A(t)x; \qquad \text{for } A(t) \in M_n, \ x \in E^n, \ x_o = x(t_o), \tag{5.2}$$

then the conditions of definition **5.1** can be simplified, since the state solution $x(t;x_o,t_o)$ initiating from x_o is linear in x_o, that is for any $\beta \in R$, $x(t;\beta x_o,t_o) = \beta x(t;x_o,t_o)$ for all $t \geq t_o$.

Definition 5.2: *Stability of linear systems*

The zero solution $x(t;x_o,t_o) = 0$ for all $t \geq t_o$ of the linear system (5.2) is said to *stable in the sense of Liapunov* if and only if for each $t_o \geq 0$ there is a finite constant $N(t_o)$ such that

$$\|x(t;x_o,t_o)\| \leq N(t_o) \|x_o\| \qquad \text{for all } t \geq t_o. \tag{5.3}$$

Geometrically definition **5.2** means that the solution to (5.2) remains for all $t \geq t_o$ in the sphere defined by (5.3).

Theorem 5.1: *Stability of linear non-autonomous systems (Willems 1970)*

The zero solution of $\dot{x} = A(t)x$, $A(t) \in M_n$, is stable in the sense of Liapunov if there is a constant N (which may depend upon t_o) such that

$$\|\Phi(t,t_o)\| \leq N(t_o) \qquad \text{for all } t \geq t_o \tag{5.4}$$

where $\Phi(t,t_o)$ is the state transition matrix of (5.2).

Consequently to check the stability of the linear system (5.2) we need to compute n solutions: namely $\phi(\cdot;o,e_i)$, for i=1,2,..,n, where $e_i \in R^n$ is a vector with zeros everywhere except for a one in the ith entry. Then if these n solutions are bounded on R_+, the matrix $\Phi(t,t_o) \triangleq \{\phi(t;t_o,e_i)\}$ is bounded for all $t \geq t_o$; hence inequality (5.4) follows. Conversely if any of these solutions is unbounded, $\Phi(t,t_o)$ is also unbounded on R and no such N can be found.

Proof: for $t \geq t_o$ and the properties of the state transition matrix (section **3.4**) we have,

$$\begin{aligned}\|x(t;t_o,x_o)\| &= \|\Phi(t,t_o)x_o\| = \|\Phi(t,o)\Phi(o,t_o)x_o\| \\ &\leq \|\Phi(t,o)\| \cdot \|\Phi(o,t_o)\| \cdot \|x_o\| \\ &\leq N\,\|\Phi(o,t_o)\| \cdot \|x_o\|\end{aligned}$$

Hence definition **5.2** can be satisfied with $N(t_o) = N\,||\Phi(o,t_o)||$. The remainder of the proof follows by contradiction: if inequality (5.4) is false, then $||\Phi(t,o)||$ is unbounded on R_+, hence at least one $\phi(t;o,I_k)$ is unbounded on R_+. Consequently inequality (5.3) cannot be satisfied for any finite $N(t_o)$ and the zero solution must be unstable in the sense of Liapunov.

For nonstationary systems it is important to distinguish between *uniform* and *nonuniform* stability properties. The zero solution to (5.1) is *uniformly stable* if the δ in definition **5.1** is independent of t_o. So for the linear system (5.2) the N of theorem **5.1** is independent of t_o, and the theorem's condition (5.4) becomes,

$$||\Phi(t,t_o)|| \quad \leq \quad N \qquad \text{for all} \quad t \geq t_o \tag{5.5}$$

and the stability of the linear system (5.2) is uniform.

Example 5.1, (Massera 1949): Consider the linear scalar system

$$\dot{x} \;=\; (4t\sin t - 2t)x, \qquad x(t_o) = x_o$$

with solution

$$x(t;x_o,t_o) \;=\; x_o \exp\{4\sin t - 4t\cos t - t^2$$

$$- 4\sin t_o + 4t_o\cos t_o + t_o^{\,2}\}.$$

We now show that whilst the equilibrium solution $x_e = 0$ is stable (in fact asympotically stable in the large) this does not imply uniform stability. Now

$$x\{(2\alpha+1)\pi;x_o,2\pi\alpha\} \;=\; x_o \exp\{(4\alpha+1)\pi(4-\pi)\}$$

is bounded for a given α, however no bound exists independently of α since $\exp\{(4\alpha+1)\pi(4-\pi)\}$ can be made as large as we wish.

Consider now the nonlinear system (5.1) in which its coefficient vector $f(t,x)$ is periodic, that is $f(t,x) = f(t+\omega,x)$, $\omega>o$ where $f:R_+ \times B \rightarrow E^n$ and $f(t,o) = 0$ for all $t \in R_+$. In this special case if the zero solution is stable it is also uniformly stable (Yoshizawa 1966), however if $f(t,x)$ is almost periodic then stability is not necessarily equivalent to uniform stability. This nonequivalence for $f(t,x) \in AP(C)$ was demonstrated by Conley and Miller (1965), who constructed an almost periodic function $f(t) \in AP(C)$ with the properties:

(i) $\int_0^T f(u)du \to \infty$ as $T \to \infty$.

(ii) There exists real sequences $\{t_n\}$, $\{T_n\}$ such that as $n \to \infty$, $T_n \to \infty$, $t_n \to \infty$ and $\int_{t_n}^{t_n+T} f(u)du < -n$ for $n = 1,2, \ldots$.

Example 5.2: Consider the scalar linear equation

$$\dot{x} = -f(t)x, \qquad f(t) \in AP(C), \tag{5.6}$$

the solutions are $x(t;x_0,t_0) = x_0 \exp(-\int_{t_0}^{t} f(u)du)$. By property (i) above, $\exp(-\int_{t_0}^{t} f(u)du) \to 0$ as $t \to \infty$ and there exists a constant N such that $\exp(-\int_{t_0}^{t} f(u)du) \le N$ (where N may depend upon t_0). Thus the zero solution of (5.6) is stable. However if we consider the solution of (5.6) through the point (t_n,x_0), $x_0 > 0$ then

$$x(t_n + T_n;x_0,t_n) = x_0 \exp(-\int_{t_n}^{t_n+T_n} f(u)du) \ge x_0 \exp(n),$$

which shows that the zero solution of (5.6) is not uniformly stable.

If the upper bound $N(t_0)\,\|x_0\|$ in equality (5.3) is replaced by a bound $B(t_0)$ which is independent of the initial state x_0, we then have *bounded* or *Lagrange stability*; if in addition this bound B is independent of t_0 the null solution is uniformly bounded, which is also equivalent to uniform stability for linear systems.

5.2 Asymptotic Stability

For the majority of practical system designs Lagrange or Liapunov stability is insufficiently precise and some kind of convergence of the state solution with increasing t to the equilibrium state is necessary, hence we say that a system is asymptotically stable if it is both convergent and stable; to be more precise we have:

Definition 5.3: *Asymptotic stability*

The zero solution of $\dot{x} = f(t,x)$ is *asymptotically stable* if it is stable and if there exists a $\delta(t_o) > 0$ such that $||x_o|| < \delta(t_o)$ implies that

$$||x(t;x_o,t_o)|| \rightarrow 0 \qquad \text{as} \qquad t \rightarrow \infty.$$

For the linear system (5.2) the above definition holds for all x_o and can be geometrically interpreted as requiring the trajectory $\{x(t;x_o,t_o)\}$ to not only remain in the sphere of radius $N(t_o)\, ||x_o||$ (by stability in the sense of Liapunov) but in addition for large t the trajectory is to come arbitrarily close to the origin of the state space. Moreover for the linear autonomous system (5.2) since $||\Phi(0,t_o)||$ is a finite number for any finite t_o and since from the properties of transition matrices $\Phi(t,t_o) = \Phi(t,0)\Phi(0,t_o)$ we can conclude that $||\Phi(t,t_o)|| \rightarrow 0$ as $t \rightarrow \infty$ for any fixed and finite t_o if and only if $||\Phi(t,0)|| \rightarrow 0$ as $t \rightarrow \infty$. The following theorem is now obvious:

Theorem 5.2: *Asymptotic stability of linear systems*

The zero solution of $\dot{x} = A(t)x$, $A(t) \in M_n$, $t \in R_+$ is asymptotically stable if there exists a finite number α (which may depend upon t_o) such that

$$||\Phi(t,t_o)|| \leq \alpha \qquad \text{for all} \qquad t \geq t_o$$

and

$$||\Phi(t,t_o)|| \rightarrow 0 \qquad \text{as} \qquad t \rightarrow \infty.$$

If in addition the bound α is independent of t_o and the above limit holds uniformly in t_o, or equivalently for some positive scalar ξ there is a $T(\xi)$ which is independent of t_o such that $||\Phi(t,t_o)|| < \xi$ for $t \geq t_o + T(\xi)$, then the linear system (5.2) is uniformly asymptotically stable.

Example 5.3: Consider the scalar system (Coppel, 1965) $\dot{x} = a(t)x$

its solution is given by $x(t) = x_o \exp(\int_{t_o}^{t} a(u)du)$, hence from definition 5.3 its zero solution is asymptotically stable if and only if $\int_o^t a(u)du \to -\infty$ as $t \to \infty$ and similarly it is uniformly stable if and only if $\int_{t_1}^{t} a(u)du$ is bounded above for each $t \geq t_1 \geq t_o$. If the parameter $a(t) = \sin\log t + \cos\log t - \beta$ for $1 < \beta < \sqrt{2}$; then $\int_o^t a(u)du = t\sin\log t - \beta t \to -\infty$, as $t \to \infty$, so the system is asymptotically stable. However selecting $t_1 = \exp(2n\pi + \theta_1)$, $t = \exp(2n\pi + \theta_2)$ where $[\theta_1,\theta_2]$ is the interval $(\theta_1 \leq \frac{\pi}{4} \leq \theta_2)$ in which $\sin\theta + \cos\theta \geq \gamma > \beta$, then as $n \to \infty$

$$\int_{t_1}^{t} a(u)du \;\geq\; (\gamma-\beta)(\exp\theta_2 - \exp\theta_1)\exp(2n\pi) \;\to\; \infty,$$

and consequently the system is not uniformly stable. It therefore follows that asymptotic stability does not imply uniform stability.

An exception to the above conclusion is for the general periodic system:

$$\dot{x} = f(t,x), \qquad x(t_o) = x_o \tag{5.7}$$

for $f: R_+ \times B = D \to E^n$, $f(t,o) = 0$ and $f(t+\omega,x) = f(t,x)$ $\omega > 0$. In as much as if the zero solution of (5.7) is asymptotically stable then it is also uniformly stable (Yoshizawa, 1966). The restriction on $f(t,o) = 0$ can be removed by utilising the following lemma for almost periodic systems with $f(t,x) \in AP(C)$:-

Lemma 5.1: *(Yoshizawa 1975)*

For the almost periodic system $\dot{x} = f(t,x)$, $f(t,x) \in AP(C)$, if for any $t_o \in R_+$ and $\xi > 0$ there exists a $\delta(t_o) > 0$ and a $T(t_o,\xi) \geq 0$ such that $||x_o|| < \delta(t_o)$ implies $||x(t;x_o,t_o)|| < \xi$ for all $t \geq t_o + T(t_o,\xi)$, then the zero solution of the system is defined on R_+.

We note that the condition $f(t,o) = 0$ is not assumed and that for the existence of the zero solution it is sufficient to assume $t_o = 0$. Now applying lemma 5.1 to the periodic system (5.7) without the condition $f(t,o) = 0$ we get:

Theorem 5.3: *Uniform asymptotic stability for periodic systems (Strauss 1969)*

For the periodic system (5.7) (without $f(t,o) = 0$) subject to the assumptions of lemma **5.1**, the system's zero solution is uniformly asymptotically stable.

Proof: The proof is relatively simple, in that the existence of the zero solution to (5.7) on R_+ follows directly from lemma **5.1**. All that needs to be shown is that the solution is unique to the right, then the solution is both asymptotically and uniformly stable (Strauss, (1969)).

Returning to the linear homogeneous equation,

$$\dot{x} = A(t)x, \quad A(t) \in M_n, \quad t \in R_+, \tag{5.8}$$

its stability properties can be summarised completely in terms of its fundamental matrix $X(t)$ as:

Theorem 5.4: *General stability conditions for linear systems*

Let $X(t)$ be a fundamental matrix of the linear system (5.8), then the zero solution to (5.8) is:

(i) Stable if and only if there exists a constant $N > 0$ such that

$$||X(t)|| \leq N \qquad \text{for all} \quad t \geq t_o \tag{5.9}$$

(ii) Uniformly stable if and only if there exists a constant $N > 0$ such that

$$||X(t)X^{-1}(s)|| \leq N \qquad \text{for} \quad t_o \leq s \leq t < \infty, \tag{5.10}$$

(iii) Asymptotically stable if and only if

$$||X(t)|| \to 0 \qquad \text{as} \quad t \to \infty, \tag{5.11}$$

(iv) Uniformly asymptotically stable if there exist positive constants N, α such that

$$||X(t)X^{-1}(s)|| \leq N\exp(-\alpha(t-s)) \qquad \text{for} \quad t_o \leq s \leq t < \infty, \tag{5.12}$$

Proof: Without loss of generality assume that $X(t_o) = I$, so that $X(t) = \Phi(t,t_o)$, the state transition matrix of (5.8). Conditions (5.9), (5.11) were established respectively in theorems **5.1** and **5.2**. The solution to (5.8) which takes on the value ζ at time s is $x(t) = X(t)X^{-1}(s)\zeta$, therefore since the solution is stable

$$\|x(t)\| \le N\|\zeta\| < \varepsilon , \tag{5.13}$$

if $N \ge \|X(t)X^{-1}(s)\|$ and $\varepsilon N^{-1} > \|\xi\|$, which holds uniformly and hence (5.10) follows. Suppose now that (5.10) does hold then for $\|\xi\| < 1$, $N > \varepsilon$ and

$$\|x(t)\| = \|X(t)X^{-1}(s)\zeta\| \le \|X(t)X^{-1}(s)\| \le N.$$

Setting $\varepsilon = N\exp(-\alpha T)$, inequality (5.13) becomes

$$\|x(t)\| \le \|X(t)X^{-1}(s)\zeta\| \le N\exp(-\alpha(t-s)) \le \varepsilon$$

for $t > s+T$. So that the null solution to (5.8) is uniformly asymptotically stable. An equivalent condition (Willems, (1970)) to condition (iv) for uniform asymptotic stability of (5.8) is if there exist positive constants N_1, α_1 such that

$$\|X(t)\| \le N_1 \exp(-\alpha_1(t-s)) \qquad \text{for} \quad \infty > t \ge s \ge t_o.$$

The above inequality is a form of *exponential stability* and is clearly a sufficient condition for uniform asymptotic stability.

We have seen in section 4.2 that the Liapunov transformation $x(t) = P(t)z(t)$ transforms the linear system (5.8) into $\dot{z} = Bz$ where $B = P^{-1}AP - P^{-1}\dot{P}$. Thus if $X(t) = P(t)Z(t)$, where $X(t)$ and $Z(t)$ are the fundamental matrices of their respective system equations, then $Z = P^{-1}X$, $X^{-1} = Z^{-1}P^{-1}$ and the boundedness of X implies boundedness of Z. If $\dot{z} = Bz$ is stable then it is also uniformly stable (that is $\|Z(t)Z^{-1}(\tau)\| \le N$) since $B \in \overline{M}_n$. Also $\|X(t)X^{-1}(\tau)\| = \|P(t)Z(t)Z^{-1}(\tau)P^{-1}(\tau)\| \le \|P(t)\| \times \|Z(t)Z^{-1}(\tau)\| . \|P^{-1}(\tau)\| \le \|P(t)\| .N. \|P^{-1}(\tau)\| \le N'$ for some N' and $0 \le \tau \le t$, and so (5.8) is uniformly stable if it is stable and reducible (or equivalently a Liapunov transformation exists).

For the special case of $A(t) = A \in \overline{M}_n$, the fundamental matrix of the linear system (5.8) is $X(t) = \exp(tA) = S\exp(tJ)S^{-1}$, where $J = \text{Diag}\{J_i\}$ is the *Jordan canonical* form of the constant matrix A and S is an invertible similarity transforming matrix of A (Desoer, 1970). From theorem **5.4** and the explicit representation of the exponential matrix in Jordan canonical form we obtain directly:

Theorem 5.5: *Stability conditions for linear time invariant systems*

The null solution to the linear time invariant system

$$\dot{x} = Ax, \qquad A \in \overline{M}_n, \qquad t \in R_+, \tag{5.14}$$

is

(i) stable if and only if every characteristic value of A has a real part not greater than zero, that is $Re(\lambda_i) \not> 0$ for all i, and those with $Re(\lambda_i) = 0$ are distinct;

(ii) asymptotically stable if and only if every characteristic value of A has negative real part (that is $Re(\lambda_i) < 0$ for all i).

Clearly stability for linear time invariant systems implies uniform stability and asymptotic stability imples uniform asymptotic stability. Theorem **5.5** is particularly significant since the stability properties of linear time invariant systems depend only upon the characteristic values of the coefficient matrix A, unlike in the general linear nonstationary system (5.8) where knowledge of the fundamental matrix and thus a complete set of independent solutions is required. The only exceptions to these restrictions is when A(t) is periodic or has certain asymptotic properties associated with large t (Rapoport, 1954).

We have shown in section **3.5** that Floquet theory allows the stability results of linear time invariant systems with coefficient matrix A to be used for linear periodic systems with the characteristic value of A being replaced by the characteristic exponents of $A(t) = A(t+\omega)$, $\omega > 0$. Thus for the periodic system

$$\dot{x} = A(t)x, \qquad A(t) = A(t+\omega) \in M_n, \qquad t \in R_+, \tag{5.15}$$

we have the following theorems directly from theorem **5.5**.

Theorem 5.6: *Uniform stability of linear periodic systems*

The null solution of (5.15) is uniformly stable if and only if A(t) has no characteristic exponent with positive real part, and if the characteristic exponents with zero real parts correspond to Jordan blocks of order 1 in the Jordan canonical form of matrix B.

The conditions of theorem **5.6** are equivalent to no characteristic multipliers of A(t) being greater than 1, and that all Jordan blocks in the Jordan canonical form of A which corresponds to characteristic multipliers of magnitude 1, are of order 1.

Theorem 5.7: *Asymptotic stability for linear periodic systems*

The null solution of (5.15) is uniformly asymptotically stable if and only if A(t) has only characteristic exponents with negative real parts, or equivalently if A(t) has only characteristic multipliers with magnitudes less than 1.

5.3 Matrix Projections and Dichotomies of Linear Systems

In this section we briefly digress to develop the theory of matrix projections and dichotomy of solutions of linear systems; so that we can derive stability conditions which are characterised only by the elements of coefficient matrices $A(t) \in M_n$ for linear homogeneous systems:

$$\dot{x} = A(t)x, \qquad x(t_o) = x_o. \tag{5.16}$$

From a practical viewpoint unstable systems are as interesting as stable systems, particularly if the factors causing instability can be identified and compensated for in system design. We consider the situation whereby the system (5.16) exhibits two kinds of solutions in state space: one which is bounded by a decaying exponential as $t \to +\infty$ and the other bounded by a growing exponential as $t \to +\infty$. Clearly the solutions continued backward in time will follow opposite bounds. So for any $t \in R$ the only bounded solution to (5.16) is the trivial one which is at the opposite extreme from having all solutions bounded. To clarify the situation define E_{oo} as the set of all points in the state space R_n which are the values for $t = 0$ of bounded solutions of (5.16); this set is clearly a non-void manifold since $0 \in E_{oo}$.

The value $t = 0$ for the definition of E_{oo} is quite arbitrary and selected for convenience. Following the solutions of (5.16) from $t = 0$ to $t_o \varepsilon R$, the set of values for $t = t_o$ of the bounded solutions of (5.16) is precisely $X(t_o)E_{oo}$, where $X(t)$ is a fundamental matrix of (5.16). By analogy with E_{oo}, define E_o as the set of all points in the state space which are values for $t = 0$ of integral curves of (5.16) which tend to zero as $t \to +\infty$; this set is a non-void linear submanifold of E_{oo} since $0 \varepsilon E_o$. In the following suppose that the state space R_n can be partitioned and defined by the direct sum of E_{oo} and another subspace E_1, which is the complement of E_{oo}. Here again the choice of $t = 0$ is arbitrary. Massera and Scheffer (1958) have shown that if the state space is the direct sum of subspaces E_{oo} and E_1, then every solution $x(t)$ of (5.16) can be expressed uniquely as:

$$x(t) = x_o(t) + x_1(t), \text{ with } x_o(t) \varepsilon X(t)E_{oo},\ x_1(t) \varepsilon X(t)E_1.$$

Alternatively, there exist projections P_o and $P_1 = I-P_o$ associated with E_{oo} and E respectively (or $X(t)P_oX^{-1}(t)$ and $X(t)(I-P_o)X^{-1}(t)$ when associated with $X(t)E_{oo}$ and $X(t)E_1$ respectively), such that

$$x(o) = P_ox(o) + (I-P_o)x(o), \tag{5.17}$$

with $P_o \cap (I-P_o) = \{0\}$ and $P_o{}^2 = P_o$ for all $x(o)$ in the state space R_n. Note that if $x(o) \varepsilon E_{oo} \cap E_1$ we have

$$x(o) = P_ox(o) = P_1x(o) = (I-P_o)x(o)$$

$$= P_o(I-P_o)x(o) = (P_o-P_o{}^2)x(o) = 0$$

so that $E_{oo} \cap E_1 = \{0\}$; in this case the projection is said to be *supplementary*. It is now obvious that every solution $x(t)$ of (5.16) can be written as

$$x(t) = X(t)(I-P_o)X^{-1}(t_o)x(t_o) + X(t)P_oX^{-1}(t_o)x(t_o) \tag{5.18}$$

This representation of the solution of (5.16) as the sum of two sets of solutions starting from E_{oo} and E_1 is unique and is referred to as a *dichotomy*. A linear transformation S leaves the subspaces E_{oo}, E_o invariant if and only if the transformation commutes with each of the projections P_o, $(I-P_o)$. If the characteristic values $\{\lambda_i\}$ of S are contained in two disjoint subsets G_{oo} and G_1, then the state space $R_n = E_{oo} \cup E$ is invariant under S, such that the characteristic values of $S \to E_{oo}$ (or $S \to E_1$) are just the elements of G_{oo} (or G_1). The corresponding projection P_o of R_n onto E_{oo} is

$$P_o = \frac{1}{(2\pi i)} \int_\Gamma (\lambda I-S)^{-1} d\lambda \tag{5.19}$$

where Γ is a closed contour in the complex characteristic value plane that contains all elements of G_{oo}, whilst all elements of G_1 lie outside Γ (Riesz and Nagy, 1955). If each $\lambda_i \in G_{oo}$ is such that $Re(\lambda_i) < \alpha$, and each $\lambda_i \in G_1$, $Re(\lambda_i) > \beta$ where $\beta > \alpha$ are scalar numbers. Then the projections P_o, $(I-P_o)$ of R_n onto E_{oo} and E_1 commute with S and there exists a constant $k > 0$ such that

$$\|\exp(St)P_o\| \le k\exp(\alpha t) \qquad \text{for} \quad t \ge 0$$

$$\|\exp(St)(I-P_o)\| \le k\exp(\beta t) \qquad \text{for} \quad t \le 0$$

An important lemma, based upon the above projection concepts, that is used to determine the stability of linear homogeneous differential equations with time-varying coefficients is:

Lemma 5.2

Let $X(t) \in M_n$ be an invertible matrix for $t \ge t_o$, and let P_o be a projection. If there exists a positive constant k such that

$$\int_{t_o}^{t} \|X(t)P_oX^{-1}(s)\| ds \le k \qquad \text{for} \quad t \ge t_o \tag{5.20}$$

then there exists a positive constant γ such that

$$\|X(t)P_o\| \le \gamma\exp(-k^{-1}t) \qquad \text{for} \quad t \ge t_o \tag{5.21}$$

Proof: It follows that Coppel (1965). Suppose that $P_o \ne 0$, and let $a(t) = \|X(t)P_o\|^{-1}$ then from the identity

$$\left[\int_{t_o}^{t} a(s)ds\right]X(t)P_o = \int_{t_o}^{t} X(t)P_oX^{-1}(s)X(s)P_oa(s)ds$$

it follows that $[a(t)]^{-1}\int_{t_o}^{t} a(s)ds \le k.$

Setting $b(t) = \int_{t_o}^{t} a(s)ds,$ therefore

$$a(t) \ge b(t_1)\exp(k^{-1}(t-t_1)) \qquad \text{for} \quad t \ge t_1$$

therefore, $\|X(t)P_o\| = [a(t)]^{-1} \le k[b(t)]^{-1}$

$$\le kb(t_1)^{-1}\exp(-k^{-1}(t-t_1)) \quad \text{for} \quad t \ge t_1$$

Then setting $\gamma \ge kb(t_1)^{-1}\exp(k^{-1}t_1)$ so large that $\|X(t)P_o\| \le \gamma\exp(-k^{-1}t)$; the lemma follows.

To return to the question of bounded solutions of the linear homogeneous system (5.16) we now make the following definition:

Definition 5.4: *Exponential dichotomy*

The homogeneous linear differential equation $\dot{x} = A(t)x$ is said to possess an *exponential dichotomy* if there exists a projection P_o and positive constants k,ℓ,α,β such that

$$\begin{aligned} \|X(t)P_oX^{-1}(t_o)\| &\le k\exp|-\alpha(t-t_o)| \qquad \text{for} \quad t \ge t_o \\ \|X(t)(I-P_o)X^{-1}(t_o)\| &\le \ell\exp|-\beta(t-t_o)| \quad \text{for} \quad t_o \ge t \end{aligned} \tag{5.22}$$

for all $t \in R$.

If in definition **5.4** the constants $\alpha=\beta=0$, then the system is said to possess an *ordinary dichotomy*. Note that for time invariant systems the existence of an exponential dichotomy is equivalent to saying that the characteristic values of the coefficient matrix A lie off the imaginary axis. It is then clear that two time-invariant systems that are related by a similarity transformation satisfy common exponential dichotomies with the same projections. This is really a result about kinematic similarity see also Chapter 4 and Appendix 2). We have already noted in theorems 5.6, 5.7 that all stability questions concerning linear homogeneous systems with periodic coefficients are given by results relating to time invariant systems with their characteristic values being replaced by the characteristic exponents of A(t). Also, like linear time invariant systems, linear periodic systems satisfy an exponential dichotomy with projection $P_o = I$, so that if the periodic system (5.15) has every characteristic exponent ρ_i such that $Re(\rho_i) < \alpha$ then there exists a constant β such that

$$\|X(t)X^{-1}(s)\| \leq \beta \exp(\alpha(t-s)) \qquad \text{for} \quad t \geq s$$

or, if $Re(\rho_i) > \alpha$, then

$$\|X(t)X^{-1}(s)\| \leq \beta \exp(\alpha(t-s)) \qquad \text{for} \quad t \leq s$$

where X(t) is the fundamental matrix of (5.15). Also by theorem 5.4 asymptotically stable time invariant systems satisfy an exponential dichotomy with $P_o = I$; the situation for other than constant or periodic coefficients A(t) is quite different as suggested by the following example due to Fink (1974):

Example 5.4

Let

$$A(t) = \begin{bmatrix} -1 + \frac{3}{2}\cos^2 t & 1 - \frac{3}{2}\cos t \sin t \\ -1 - \frac{3}{2}\sin t \cos t & -1 + \frac{3}{2}\sin^2 t \end{bmatrix}$$

The characteristic values of A(t) are given by solution of $\text{Det}(\lambda(t)I - A(t)) = 0$ for all t as the *constants* $\frac{1}{4}(-1 \pm i\sqrt{7})$. Yet the system $\dot{x} = A(t)x$ possesses solutions of the form $(-\cos t, \sin t)\exp(\frac{1}{2}t)$ whose norm $\to +\infty$ as $t \to \infty$.

Fortunately there are several known conditions (Coppel, 1967 (2)) that give exponential dichotomy in the time-dependent case; the following will be stated without proof:

Theorem 5.8: *Exponential dichotomy conditions (Coppel 1967(2))*

Let $A(t) \varepsilon M_n$ possess m characteristic values λ_i such that $Re(\lambda_i) \le -\alpha$, $\alpha > 0$ and (n-m) characteristic values with $Re(\lambda_i) \ge \beta > 0$. Then for $\min(\alpha,\beta) > \eta > 0$ there is a constant $\delta = \delta(N,\alpha+\beta,\eta)$, where $||A(t)|| \le N$, such that if $||\dot{A}(t)|| < \delta$ then the system $\dot{x} = A(t)x$ satisfies an exponential dichotomy (5.22) with

$$P_o = \begin{bmatrix} I_m & 0 \\ 0 & 0 \end{bmatrix}$$

and k,ℓ depending only on N, $\alpha+\beta$ and η.

This theorem indicates that $\dot{A}(t)$ was too large in example 5.4 and illustrates the dangers of determining system stability based only on the characteristic values of time dependent coefficients A(t). However, rather conservative sufficient conditions for exponential dichotomy of (5.16) based explicitly upon the diagonal dominance of the coefficient matrix A(t) can be developed from Gershgorin's theorem. This theorem locates the characteristic values $\lambda_1(A), \dots \lambda_n(A)$, of the matrix A in the union of circles in the complex plane centred along the diagonal elements $\{a_{ii}(t)\}$ of the coefficient matrix A(t) (Gantmacher 1959, and Chapter One). The radii r_i of these circles are linear functions of the off-diagonal elements of A(t), and three types of diagonal dominance can be identified.

Consider the linear homogeneous system,

$$\dot{x} = A(t)x, \qquad A(t) \varepsilon M_n \tag{5.23}$$

If the coefficient matrix A(t) is such that $|Re(a_{ii}(t))| \ge \sum_{\substack{j=1 \\ j\neq i}}^{n} |a_{ji}(t)| + \xi$, for all i and any $\xi > 0$, then the matrix A is said to be *column dominant* with Gershgorin circles $r_i = \sum_{\substack{j=1 \\ j\neq i}}^{n} |a_{ji}(t)|$. If in addition we use the ℓ^1 norm for the state vector of (5.23) then the measure of the matrix operator is given by (see section 1.5),

$$\mu_1(A) = \max_j \{\text{Re}(a_{ii}(t)) + \sum_{\substack{j=1 \\ j\neq i}}^{n} |a_{ji}(t)|\} < -\xi$$

if $a_{ii}(t) < 0$ for all i. Utilising Coppel's inequality for norm of x based upon $\mu(A)$ (see section 1.5), we have

$$||x(t)|| \leq ||x(t_o)|| \exp(-\xi(t-t_o)) \qquad \text{for} \quad t \geq t_o$$

Hence the system (5.23) has an exponential dichotomy with projection $P_o = I$ if A(t) is column-dominant with $a_{ii}(t) < 0$ for all i. In addition the system (5.23) is exponentially stable by the proof of theorem 5.4.

If now the elements of coefficient matrix A(t) are such that $|\text{Re}(a_{ii}(t))| \geq \sum_{\substack{j=1 \\ j\neq i}}^{n} |a_{ij}(t)| + \xi$, for all i and $\xi > 0$, then the matrix A(t) is said to be *row-dominant* with circles $r_i = \sum_{\substack{j=1 \\ j\neq i}}^{n} |a_{ij}(t)|$. If we now select the ℓ^∞ norm for x, the measure of the matrix operator for this norm is $\mu_\infty(A) = \max_i\{\text{Re}(a_{ii}(t)) + \sum_{\substack{j=1 \\ j\neq i}}^{n} |a_{ij}(t)|\}$. Thus if A(t) is row-dominant with $k(k \leq n)$ diagonal elements $a_{ii}(t) < 0$ for all t, then $\mu_\infty(A) \leq -\xi$, and if the remaining (n-k) $a_{ii}(t) > 0$ for all t then $\mu_\infty(A) \leq \xi$. Hence the system (5.23) has an exponential dichotomy, since there are k independent solutions $x_1, \ldots x_k$ such that $x \in \text{span}(x_1, \ldots x_k)$ with $||x||$ strictly increasing and $||x(t_o)|| \exp(\xi(t-t_o)) \leq ||x(t)||$ for $t \geq t_o$, and (n-k) independent solutions $(x_{k+1}, \ldots x_n)$ such that $x \in \text{span}(x_{k+1}, \ldots x_n)$ with $||x||$ strictly decreasing and $||x(t)|| \leq ||x(t_o)|| \exp(-\xi(t-t_o))$ (Lazer, 1971).

A third type of diagonal dominance of A(t) exists which is called *mean dominance*, if, $|\mathrm{Re}(a_{ii}(t))| \geq \frac{1}{2}\{\sum_{\substack{j=1\\j\neq i}}^{n} |a_{ij}(t)| + \sum_{\substack{j=1\\j\neq i}}^{n} |a_{ji}(t)|\} + \xi$, for all i and any $\xi > 0$. In this case the Gershgorin circles are of radii $r_i = \sum_{\substack{j=1\\j\neq i}}^{n} (|a_{ij}(t)| + |a_{ji}(t)|)$.

Then if we use the ℓ^2 norm for x then the measure of A is given by $\mu_2(A) = \max_i \left\{\lambda_i \frac{(A+A^*)}{2}\right\} \leq -\xi$ if $a_{ii}(t) < 0$ for all t and i. Applying Coppel's inequality (1.32) to this measure gives the inequality

$$\|x(t)\| \leq \|x(t_o)\| \exp(-\xi(t-t_o)) \qquad \text{for} \quad t \geq t_o,$$

which is an exponential dichotomy with $P_o = I$ and a condition for exponential stability for system (5.23). Indeed for any diagonal dominant matrix A(t) with $a_{ii}(t) < 0$ for all t and i, the above inequality holds for any norm since all norms are equivalent. In conclusion we see that $|\mathrm{Re}(a_{ii}(t))| > r_i$ is a sufficient condition for exponential dichotomy of the linear homogeneous system (5.23).

5.4 Asymptotic Characteristic Value Stability Theory

A less conservative but more restrictive theory for the stability of linear nonstationary systems based upon the asymptotic properties of the characteristic values $\{\lambda(t)\}$ of A(t) can be developed via matrix projection theory. Consider the linear homogeneous system (5.23) but with the additional condition that $\lim_{t\to\infty} A(t) = A_\infty \in \overline{M}_n$. It will be shown in the sequel that if λ_∞ is a distinct characteristic value of A_∞ then A(t) has a unique characteristic value (exponent) $\lambda(t)$ in the neighbourhood of λ_∞ such that $\lim_{t\to\infty} \lambda(t) = \lambda_\infty$.

Let Γ be a closed contour in the complex characteristic value plane which includes λ_∞, but does not include any other characteristic value of A_∞, then there exists a projection

$$P_o = \frac{1}{(2\pi i)} \int_\Gamma (\lambda I - A_\infty)^{-1} d\lambda,$$

which commutes with A_∞ (see also equation (5.19)). For all points λ on Γ $\|\lambda I - A_\infty\| \geq \xi > 0$. If t_o is chosen so large that $\|A(t) - A_\infty\| \leq \frac{\xi}{2}$ for $t \geq t_o$ then $\|\lambda I - A(t)\| \geq \frac{\xi}{2}$ and

$$P(t) = \frac{1}{(2\pi i)} \int_\Gamma (\lambda I - A(t))^{-1} d\lambda \tag{5.24}$$

is defined as a projection which commutes with $A(t)$. Setting

$$\begin{aligned} S(t) &= (I - P(t))(I - P_o) + P(t)P_o \\ &= I + (P(t) - P_o)(2P_o - I), \end{aligned} \tag{5.25}$$

so that $P(t)S(t) = P(t)P_o = S(t)P_o$ and $\lim_{t\to\infty} S(t) = I$, then $S^{-1}(t)$ exists for all large t and $P(t) = S(t)P_o S^{-1}(t)$. It then follows that the characteristic vector $\Lambda_t = S(t)\Lambda_\infty$, belonging to the characteristic value $\lambda(t)$ of $A(t)$ inside Γ. Thus

$$S^{-1}(t)A(t)S(t)\Lambda_\infty = \lambda(t)\Lambda_\infty \tag{5.26}$$

where Λ_∞ is the characteristic vector of matrix A_∞ associated with λ_∞ inside Γ. Then if $A(t)$ is continuous and differentiable r times then by equations (5.24-26) so is $P(t)$, $S(t)$ and $\lambda(t)$.

Also if $\int_t^\infty \|\dot{A}(s)\| ds < \infty$, then differentiating (5.24)

$$\begin{aligned} \dot{P}(t) &\leq (2\pi i)^{-1} \int_\Gamma (\lambda I - A(t))^{-1} \dot{A}(t) (\lambda I - A(t))^{-1} d\lambda \\ &\leq (2\pi)^{-1} \|\dot{A}(t)\| \int_\Gamma \|(\lambda I - A(t))^{-1}\|^2 d\lambda \\ &\leq \pi^{-1} \|\dot{A}(t)\| \int_\Gamma \|(\lambda I - A_\infty)^{-1}\|^2 d\lambda, \end{aligned}$$

hence also $\int_t^\infty ||\dot{P}(s)||ds < \infty$ and $\int_t^\infty |\lambda(s)|ds < \infty$ for large t.

Similarly from (5.25), $\dot{S}(t) = \dot{P}(t)(2P_o - I)$ then $||(S^{-1})'|| \leq$

$||S^{-1}||^2 . ||\dot{S}|| \to ||\dot{S}||$ for large t, and hence $\int_t^\infty ||\dot{S}(s)||ds < \infty$.

We are now able to use the transformation matrix S(t) to diagonalize A(t) into a Jordan type matrix. We know that for λ_∞ a distinct characteristic value of A_∞, there exists a constant similarity type transformation matrix S such that

$$D_\infty = S^{-1}A_\infty S = \begin{bmatrix} \lambda_\infty & 0 \\ 0 & D_{\infty-1} \end{bmatrix}$$

Moreover $D(t) = S^{-1}A(t)S$ are linear combinations of the elements of A(t) with constant coefficients and are such that

$\int_t^\infty ||\dot{D}(s)||ds < \infty$; we may then assume that

$$A_\infty = \begin{bmatrix} \lambda_\infty & 0 \\ 0 & A_{\infty-1} \end{bmatrix}$$ and that S(t) satisfies:

Lemma 5.3: *(Coppel 1965)*

Let $A(t) \in M_n$ such that $\int_0^\infty ||\dot{A}(s)||ds < \infty$ and let λ_∞ be a distinct characteristic value of $A_\infty = \lim_{t\to\infty} A(t)$. Then there exists for large t an invertible matrix $S(t) \in M_n$ such that

$$[S(\infty)]^{-1} = [\lim_{t\to\infty} S(t)]^{-1} \quad \text{exists},$$

$$\int_0^\infty ||\dot{S}(s)||ds < \infty, \quad \text{and}$$

$$S^{-1}(t)A(t)S(t) = \begin{bmatrix} \lambda(t) & 0 \\ 0 & A_{-1}(t) \end{bmatrix}$$

where $\lambda(t)$ is the characteristic root of $A(t)$ such that $\lim_{t\to\infty} \lambda(t) = \lambda_\infty$ and $A_{-1}(t) \in M_{n-1}$.

This lemma is now used to establish a theorem similar to theorem **5.6** for the general class of coefficient matrices $A(t) \in M_n$ which are of bounded variation onthe interval $[t_0,\infty)$.

Theorem 5.9: *Asymptotic uniform stability (Conti, 1955)*

The linear homogeneous equation $\dot{x} = A(t)x$ with $A(t) \in M_n$ and $\int_{t_0}^{\infty} \|\dot{A}(s)\| ds < \infty$ has uniformly stable solutions if and only if:

(i) the characteristic values of $A(t)$ have non-negative real parts for $t \geq t_0$;

(ii) the characteristic roots of $A_\infty = \lim_{t\to\infty} A(t)$, whose real parts are zero, are distinct.

Proof: Let λ_i and $\lambda_i(t)$ be respectively the characteristic values of A_∞ and $A(t)$ and are related by $\lambda_i = \lim_{t\to\infty} \lambda_i(t)$, $(i = 1,2,\ldots n)$.

By lemma **5.3** a $S(t) \in M_n$ exists such that

$$S^{-1}(t)A(t)S(t) = \begin{bmatrix} \Lambda(t) & 0 \\ & \\ 0 & A_{-m}(t) \end{bmatrix},$$

where $\Lambda(t) = \text{Diag}(\lambda_1(t),\ldots,\lambda_m(t))$. So by making this time a varying transformation $y(t) = S^{-1}(t)x(t)$ to the system equation

$$\dot{x} = A(t)x, \qquad A(t) \in M_n, \tag{5.27}$$

it is uniformly stable at the same time as the system

$$\dot{y} = (S^{-1}(t)A(t)S(t) - S^{-1}(t)\dot{S}(t))y;$$

which in turn since $\int_0^{\infty} \|S^{-1}(s)\dot{S}(s)\| ds < \infty$ is uniformly stable at the same time as the system,

$$y = S^{-1}(t)A(t)S(t)y. \tag{5.28}$$

Let $Y(t)$ be the principal fundamental matrix of (5.28) (i.e. $Y(t_o) = I$) then

$$Y(t) = \begin{bmatrix} \Sigma(t) & 0 \\ 0 & Y_{-m}(t) \end{bmatrix}$$

where $\Sigma(t) = \mathrm{Diag}\{\exp(\int_{t_o}^{t} \lambda_1(s)ds, \ldots, \exp(\int_{t_o}^{t} \lambda_m(s)ds)\}$.

So if $\mathrm{Re}\{\lambda_i(t)\} \le 0$ $(i = 1,2,\ldots,m)$ then by theorem 5.4 part (ii), the system (5.27) will be uniformly stable over $[t,\infty)$ if and only if for some $N > 0$,

$$\|Y_{-m}(t)Y_{-m}^{-1}(s)\| \le N, \qquad \text{for} \quad \infty > t \ge s \ge t_o.$$

This inequality is satisfied since $A_{-m}(\infty)$ (and by similarity $Y_{-m}(t)$) has characteristic values with negative real parts, and hence $\|X(t)X^{-1}(s)\|$ is bounded for $\infty > t \ge s \ge t_o$ and so (5.27) is uniformly stable by theorem 5.4.

A stronger result on uniform asymptotic stability of (5.27) can be derived if $A(t) \in M_n$ and $\|\dot{A}(t)\|$ is small (or equivalently satisfies a Lipschitz condition in t):

Theorem 5.10: *Uniform asymptotic stability (Lyascenko, 1954)*

Let $A(t) \in M_n$ such that $\|\exp(\tau A(t))\| \le N\exp(-\alpha\tau)$ for $\tau \ge 0$, $t \ge t_o$ with $N > 1$ and $\alpha > 0$, and $\|A(t_1)-A(t_2)\| \le k(t_2-t_1)$ for $t_1 \ge t_o$, $t_2 \ge t_o$ where $k < \alpha^2(N\log N)^{-1}$. If $X(t)$ is the fundamental matrix for

$$\dot{x} = A(t)x, \tag{5.29}$$

then

$$\|X(t)X^{-1}(s)\| \le N^2\exp(-\beta(t-s)) \qquad \text{for} \quad t \ge s \ge t_o,$$

where $\beta = \alpha - (kN\log N)^{\frac{1}{2}} > 0$, and the system (5.29) is uniformly asymptotically stable.

Proof: For any $\tau \geq t_o$, (5.29) can be rewritten as

$$\dot{x} = A(\tau)x(t) + (A(t) - A(\tau))x(t)$$

hence,

$$x(t) = X(t)x(s) + \int_s^t X(t)X^{-1}(u)(A(u)-A(\tau))x(u)du.$$

Therefore

$$\|x(t)\| \leq N\exp(-\alpha(t-s))\ \|x(s)\| + N\int_s^t \exp(-\alpha(t-u)) \times$$
$$\|A(u)-A(\tau)\| \cdot \|x(u)\| du$$

Applying the Gronwell-Bellman lemma to the above inequality gives

$$\|x(t)\| \leq N\exp(-\alpha(t-s))\exp\{N\int_s^t \|A(u)-A(\tau)\| du\} \cdot \|x(s)\|$$

for $t \geq s$. Setting $t = s+\gamma$, $\tau = s+\frac{\gamma}{2}$ and utilising the Lipschitz condition for $\gamma \geq 0$,

$$\|x(s+\gamma)\| \leq N\exp(-\alpha\gamma)\exp\left(\frac{kN\gamma^2}{4}\right) \|x(s)\| .$$

So putting $\gamma = 2[(kN)^{-1}\log N]^{\frac{1}{2}}$ or $N = \exp\left(\frac{kN\gamma^2}{4}\right)$ in the above,

$$\|x(s+\gamma)\| \leq \exp(-\beta\gamma)\ \|x(s)\|$$

$$\leq N^2\exp(-\alpha\gamma)\ \|x(s)\| , \tag{5.30}$$

where $\beta = \alpha-(kN\log N)^{\frac{1}{2}}$. Clearly $\alpha \geq \beta$ and the condition

$$\|x(t)\| \leq N^2\exp(-\beta(t-s))\ \|x(s)\|$$

follows from (5.30). Finally substituting $x(t) = X(t)X^{-1}(s)\xi$ in the above inequality for ξ an arbitrary vector

$$\|X(t)X^{-1}(s)\| \leq N^2\exp(-\beta(t-s)) \qquad \text{for} \quad t \geq s \geq t_o.$$

We note from theorem 5.4 part (iv) that this is the necessary and sufficient condition for the uniform asymptotic stability of (5.29). There is an obvious connection between theorem 5.10

and theorem **5.8**, since on setting $P_o = I_n$ in theorem **5.8** all the characteristic values of A(t) have negative real parts and given that $|\dot{A}(t)| < \delta$ we satisfy the conditions of theorem **5.10**.

Theorems **5.8-5.10** demonstrate that stability conditions for linear nonstationary homogeneous systems can be expressed in terms of the asymptotic properties of the characteristic values of the coefficient matrix A(t) and the fundamental matrix of the system equations; the question now arises can general nonlinear system equations of the form of (5.1) be treated in like manner? Suppose that the nonlinear vector f(t,x) can be decomposed by a process of linearisation into $f(t,x) = A(t)x + g(t,x)$, where $A(t) \varepsilon M_n$, $f: R_+ \times B = D \to E^n$ for $B = \{x : x \varepsilon E^n, \|x\| < \alpha, \alpha > 0\}$. The nonlinear equation

$$\dot{x} = A(t)x + g(t,x), \qquad x(t_o) = x_o \tag{5.31}$$

has a solution through x_o, which is not necessarily unique unless a condition such as $\|g(t,x)\| \leq \delta(t)\|x\|$ where $\delta \not\to 0$ and integrable (Curtain and Pritchard, 1977) is imposed. The linearised element of (5.1) (or (5.31)) is the homogeneous linear equation

$$\dot{x} = A(t)x, \qquad A(t) \varepsilon M_n, \tag{5.32}$$

whose fundamental matrix is X(t) with $X(t_o) = I$. We now establish results that relate the stability of the linear system (5.32) to the stability of the null solution to (5.31).

Theorem 5.11: *(Caligo, 1940; Conti, 1955)*

If g(t,x) satisfies the inequality

$$\|g(t,x)\| \leq \delta(t)\|x\|, \tag{5.33}$$

where $\delta(t)$ is non-negative and integrable over $[t_o,\infty)$ and if there exists a $N > 0$ such that

$$\|X(t)X^{-1}(s)\| \leq N, \qquad \text{for} \quad \infty > t \geq s \geq t_o. \tag{5.34}$$

Then there exists a solution to (5.31) which satisfies

$$\|x(t)\| \leq \beta\|x(s)\| \qquad \text{for all} \quad t \geq s \geq t_o \tag{5.35}$$

where the constant $\beta > 0$ is such that $\|x(s)\| < \beta^{-1}\alpha$. If in addition $\lim_{t\to\infty} X(t) = 0$ then $\lim_{t\to\infty} x(t) = 0$.

Corollary I: If the linear equation (5.32) is uniformly (and asymptotically) stable, and if condition (5.33) holds, then the null solution to (5.31) is uniformly (and asymptotically) stable.

Corollary II: If $g(t,x)$ is such that $g(t,x) = B(t)x$, where $B(t) \in M_n$ and integrable over $[t_o,\infty)$, then the linear system

$$\dot{x} = (A(t) + B(t))x, \qquad \text{for} \quad A,B \in M_n \tag{5.36}$$

is uniformly (and asymptotically) stable if and only if the linear system (5.32) is uniformly (and asymptotically) stable.

Proof: Follows similarly to that of theorem **5.10**. The solution of (5.31) is

$$x(t) = X(t)X^{-1}(s)x(s) + \int_s^t X(t)X^{-1}(u)g(u,x(u))du. \tag{5.37}$$

Taking norms and utilising the properties (5.34), (5.35) we get on applying the Gronwell-Bellman lemma to (5.37),

$$\begin{aligned}
\|x(t)\| &\le N\ \|x(s)\| + N\int_s^t \delta(u)\ \|x(u)\|du \\
&\le N\|x(s)\|\exp\{N\int_s^t \delta(u)du\} \\
&\le \beta\|x(s)\| \qquad \text{for} \quad t \ge s,
\end{aligned}$$

where $\beta = N\exp\{N\int_s^t \delta(u)du\}$. Finally setting $s = t_o$ in (5.37) so that $X(s) = X(t_o) = I$ and taking norms yields

$$\|x(t)\| \le \|X(t)\|\ \|x_o\| + \|\int_{t_o}^t X(t)\ X^{-1}(u)\delta(u)\|x(u)\|du\|$$

$$\leq \quad \|X(t)\| \; \|x_o\| + N\beta\|x_o\| \int_{t_o}^{t} \delta(u)du. \tag{5.38}$$

So if $\lim_{t\to\infty} X(t) = 0$, it follows directly from inequality (5.38) that $\lim_{t\to\infty} x(t) = 0$, since $\delta(s)$ is integrable by definition. Corollaries I, II follow immediately from condition (5.34) and inequality (5.38) on application of theorem **5.4**.

By further requiring that the left hand side of inequality (5.34) be integrable over $[t_o,t]$, then $X(t)$ is bounded and in particular $\lim_{t\to\infty} X(t) = 0$ (see lemma **5.2**); also if $\delta(t)$ is independent of t a simple condition for asymptotic stability of (5.31) follows:-

Theorem 5.12: *Asymptotic stability of systems with small nonlinearities*

If the fundamental matrix $X(t)$ of the linear homogeneous system (5.32) is such that

$$\int_{t_o}^{t} \|X(t)X^{-1}(s)\|ds \leq N, \qquad \text{for} \quad t \geq t_o \quad \text{and} \quad N > 0, \tag{5.39}$$

and $g(t,x)$ satisfies the inequality $\|g(t,x)\| \leq \delta\|x\|$ for $\delta < N^{-1}$, then the null solution to (5.31) is asymptotically stable.

Proof: follows directly from application of lemma **5.2** to inequality (5.39). Also if the $\delta > 0$ is sufficiently small for $A(t)$ a periodic or time invariant coefficient matrix, the asymptotic stability of the linear system (5.32) implies asymptotic stability of the null solution to the nonlinear system (5.31) since inequality (5.39) is automatically satisfied if (5.32) is asymptotically stable for $A(t)$ periodic or time invariant. In addition the asymptotic stability of systems (5.31) and (5.32) for $A(t)$ periodic or time invariant is uniform. A result for uniform asymptotic stability of the nonlinear system (5.31) for a more general class of coefficient matrices $A(t) \in M_n$ can be established by condition (iv) of theorem 5.4:-

Theorem 5.13

If the fundamental matrix of (5.32) satisfies the inequality

$$||X(t)X^{-1}(s)|| \leq N\exp(-\beta(t-s)), \quad \text{for } \infty > t \geq s \geq t_0 \qquad (5.40)$$

for positive constants N, β, and if $||g(t,x)|| \leq \delta||x||$ for $\delta < N^{-1}\beta$. Then every solution of (5.31) is defined for all $t \geq t_0$ and satisfies

$$||x(t)|| \leq N\exp(-\gamma(t-s))\,||x(s)||, \quad \text{for } t \geq s \geq t_0, \qquad (5.41)$$

and $||x_0|| < N^{-1}\alpha$, where $\gamma = \beta - \delta N > 0$.

Corollary I: If the linear system (4.32) is uniformly asymptotically stable and if δ is sufficiently small, then the zero solution of (5.31) is also uniformly asymptotically stable.

Corollary II: If the linear system (5.32) is uniformly asymptotically stable and $g(t,x) = B(t)x$ for $B(t) \in M_n$ and $\lim_{t\to\infty} B(t) = 0$, then the system $\dot{x} = (A(t)+B(t))x$ is also uniformly asymptotically stable.

Proof: By theorem 5.4 (iv), inequality (5.40) is a necessary and sufficient condition for uniform asymptotic stability of the linear system (5.32). By taking norms of the solution (5.37) and substituting inequality (5.40) into the result, the Gronwell-Bellman lemma yields inequality (5.41) directly. And since inequality (5.41) is the condition for exponential asymptotic stability for x(t) (Yoshizawa, 1966), it is sufficient to imply uniform asymptotic stability for x(t) and corollaries **I, II** therefore follow directly. For a collection of examples of corollary **II** see Cesari (1940), in which B(t) is not necessarily stable but is integrable over R_+.

Example 5.5: This example, due to Perron (1930) demonstrates that although the linear system $\dot{x} = A(t)x$ is asymptotically stable, it is not necessarily uniformly stable and a linear system $\dot{x} = (A(t)+B(t))x$ with $B(t) \in M_n$ and integrable over R_+

with $\lim_{t\to\infty} B(t) = 0$ can be unstable.

Let

$$A(t) = \begin{bmatrix} -\alpha & 0 \\ 0 & \sin(\log t) + \cos(\log t) - 2\alpha \end{bmatrix}$$

with $1 < 2\alpha < 1 + \exp(-\pi)$. Whence the solution to (5.32) is

$$x(t) = \begin{bmatrix} x_1(0)\exp(-\alpha t) \\ x_2(0)\exp(t\sin(\log t) - 2\alpha t \end{bmatrix}$$

which tends exponentially to zero as $t\to\infty$. But if we take

$$B(t) = \begin{bmatrix} 0 & 0 \\ \exp(-\alpha t) & 0 \end{bmatrix}$$

then the solutions to $\dot{x} = (A(t)+B(t))x$ are

$$x(t) = \begin{bmatrix} x_1(o)\exp(-\alpha t) \\ \exp(t\sin(\log t)-2\alpha t)[x_2(o)+x_1(o)\int_{t_o}^{t}\exp(-u\sin(\log u))du] \end{bmatrix}$$

Select a β such that $0 < \beta < \frac{\pi}{2}$ and then $\cos\beta > (2\alpha-1)\exp\pi$. So if $t_r = \exp(2r-\frac{1}{2})\pi$, $r=1,2,\ldots,$ then for $t_r \le s \le t_r\exp\beta$,

$$\int_o^{t_r\exp\pi} \exp(-u\sin(\log u))du > \int_{t_r}^{t_r\exp\beta} \exp(u\cos\beta)du$$

$$> t_r(\exp\beta-1)\exp(t_r\cos\beta)$$

since $\sin(\log u) \le -\cos\beta$. Also since $\sin(\log(t_r\exp\pi)) = 1$ it then follows as $r\to\infty$ that

$$|x_2(t_r\exp\pi)| \ge |x_1(o)|t_r(\exp\beta-1)\exp(\gamma t_r) \to \infty$$

if $x_1(0) \neq 0$, since $\gamma = (1-2\alpha)\exp\pi + \cos\beta > 0$.

Example 5.6: consider the scalar almost periodic differential equation

$$\dot{x} = -(a(t)-b(t))x, \quad \text{for } a(t) \in AP(C) \text{ and } t \in R_+, \quad (5.42)$$

where $a(t) = \sum_{k=1}^{\infty} k^{-\frac{3}{2}} \sin(\pi t k^{-1})$ and

$$b(t) = \begin{cases} \delta & \text{for } 0 \le t \le 1 \\ \delta t^{-\frac{1}{4}} & \text{for } t \ge 1 \text{ and } \delta \ge 0 \end{cases}$$

Clearly $|b(t)| \le \delta$ and $\lim_{t\to\infty} b(t) = 0$. The solution to (5.42) is $x(t) = x_0 \exp\left\{\int_0^t (b(s)-a(s))ds\right\}$. For the given a(t), positive constants α, β exist such that for $t \ge 1$

$$\alpha t^{\frac{1}{2}} \ge \int_0^t a(s)ds \ge \beta t^{\frac{1}{2}}.$$

so that $\dot{x} = -a(t)x$ is asymptotically stable, but since $\int_0^t b(s)ds = \frac{1}{3}(4t^{\frac{3}{4}}-1)$ the system (5.42) is unstable for every $\delta > 0$.

5.5 Stability in the Large

The stability properties of the previous sections were all local, that is there exists a closed domain in state space that includes the equilibrium state such that all solutions initiating in that region are stable or asymptotically stable. In the case of asymptotic stability, where there is convergence to the equilibrium state, the region of validity of convergence is called the *domain of attraction*. Should the domain of attraction include the whole state space we then have *global* or *stability in the large*.

Consider the general nonlinear system

$$\dot{x} = f(t,x), \quad f(t,0) = 0 \quad \text{and} \quad f:R_+ \times B = D \to E^n, \qquad (5.43)$$

Definition 5.5: *Asymptotic stability in the large*

The zero solution of (5.43) is *asymptotically stable in the large* or *globally stable* if it is stable and every solution of (5.43) tends to zero as $t \to \infty$.

Definition 5.6: *Exponential asymptotic stability in the large*

The zero solution of (5.43) is exponentially asymptotically stable in the large if there exists a $\alpha > 0$ and for any $\beta > 0$ there exists a $N(\beta) > 0$ such that if $||x_0|| \le \beta$ then

$$||x(t;x_0,t_0)|| \le N(\beta)\exp(-\alpha(t-t_0))||x_0|| \quad \text{for all} \quad t \ge t_0.$$

Definition 5.7: *Weakly uniformly asymptotic stability in the large*

The general solution $y(t)$ of (5.43) defined on R_+ is said to be *weakly uniformly asymptotically stable in the large* if it is uniformly stable and if for every $t_0 \in R_+$ and every x_0 defined on R_+ we have

$$\lim_{t\to\infty} ||x(t;x_0,t_0) - y(t)|| = 0$$

If $f(t,x) = f(t+\omega,x)$, $\omega > 0$ is periodic then weakly uniformly asymptotic stability in the large is equivalent to uniform asymptotic stability in the large (Yoshizawa, 1975). However as we shall see in the following example (Seifert, 1968) this equivalence is not the case for almost periodic functions $f(t,x) \in AP(C)$.

Example 5.7: Consider the scalar system,

$$\dot{x} = \begin{cases} x, & \text{for} \quad 0 \le x \le 1 \\ -1 + (1-2f(t))(x-1), & \text{for} \quad 1 < x \le 2 \\ -f(t)x, & \text{for} \quad 2 < x \end{cases} \qquad (5.44)$$

where $f(t,x) = -f(t,-x)$ and $f(t) \varepsilon AP(C)$ is the almost periodic function constructed by Conley and Miller (1965) and discussed in example **5.2.** The zero solution to (5.44) is uniformly asymptotically stable. Assume that $|f(t)| < 1$ then $f(t,x) \leq -f(t)x$. Comparing the solution of $\dot{x} = -f(t)x$ (see example 5.2) with that of (5.44) we see that every solution of (5.44) tends to zero as $t \to \infty$, and thus the zero solution to (5.44) is weakly uniformly stable in the large. Considering the solution of (5.6) through (t_n, x_o) we showed that the solution to (5.6) is not uniformly bounded and hence solutions to (5.44) are not uniformly bounded and the zero solution of (5.44) is not uniformly asymptotically stable.

Finally if we now consider the linear nonstantionary system

$$\dot{x} = A(t)x, \qquad A(t) \varepsilon M_n, \tag{5.45}$$

a variety of stability conditions are equivalent and are given without proof (Yoshizawa, 1975):-

Theorem 5.14

If the zero solution of the linear system (5.45) is asymptotically stable it is asymptotically stable in the large. Moreover if the zero solution of (5.45) is uniformly asymptotically stable it is exponentially asymptotically stable in the large and the $N(\beta)$ of definition 5.7 is independent of β.

Theorem 5.15

For the linear system (5.45)

(i) Asymptotic stability and ultimate boundedness are equivalent.

(ii) Uniform asymptotic stability in the large and uniform ultimate boundedness are equivalent.

(iii) If $A(t)$ is periodic in t, asymptotic stability in the large implies uniform asymptotic stability in the large.

5.6 Total Stability and Stability under Disturbances

Consider the general nonlinear system

$$\dot{x} = f(t,x), \tag{5.46}$$

with $f: R \times L \to E^n$ where $L < B = \{x : x \in E^n, \|x\| < \alpha, \alpha > 0\}$.

Definition 5.8: *Total stability*

Let y(t) be a solution to (5.46) such that $\|y(t)\| \leq \beta$ for all $t \geq 0$. Then the solution y(t) is said to be *totally stable* if for any $t_o \geq 0$ and any $\xi \geq 0$ there exists a $\delta(\xi) > 0$ such that if g(t) is any continuous function on $[t_o,\infty)$ with $\|g(t)\| < \delta(\xi)$ for all $t \geq t_o$ and if $z_o \in L$ satisfies $\|y(t_o) - z_o\| < \delta(\xi)$ then any solution z(t) through (t_o, z_o) of the system

$$\dot{z} = f(t,z) + g(t) \tag{5.47}$$

satisfies $\|y(t) - z(t)\| < \xi$ for all $t \geq t_o$.

If we restrict the function f(t,x) such that it satisfies a Lipschitz condition in x with f(t,o) = 0, then if the null solution to (5.46) is uniformly asymptotically stable it is also totally stable. Clearly total stability implies uniform stability but the converse is not in general true. Moreover total stability does not necessarily imply asymptotic stability; an exception to this if $f(t,x) = A(t)x$ with $A(t) \in M_n$ on R_+. In which case we have:

Theorem 5.16: *(Massera, 1958)*

If the null solution to (5.45) is totally stable then it is uniformly asymptotically stable and exponentially stable in the large.

Proof: From definition **5.8**, if the null solution is totally stable then there exists a $\delta > 0$ such that if $\|z_o\| < \delta$ the solution $z(t;z_o,t_o)$ of

$$\dot{z} = A(t)z + \delta z$$

satisfies $\|z(t;z_o,t)\| < 1$. But the solution of the above differential equation and (5.45) are related by

$$z(t;z_o,t_o) = x(t;z_o,t_o)\exp(\delta(t-t_o)) \qquad \text{for} \quad t \geq t_o,$$

then $\|x(t;z_o,t_o)\| < \exp(-\delta(t-t_o))$. Consequently by theorem 5.4 (iv) and theorem 5.14 the null solution of (5.45) is both uniformly asymptotically stable and exponentially asymptotically stable in the large.

We shall now relate the concept of total stability to Σ-stability and stability under disturbances for almost periodic systems. Consider the system (5.46) but with $f(t,x) \varepsilon AP(C)$ almost periodic in t uniformly for $x \varepsilon L$ and for all $t \geq 0$. Let Q be a compact set such that $Q \subset L \subset B$, and $y(t) \varepsilon Q$ for all $t \geq 0$. For $g \varepsilon H(f)$ (the hull of f - see section 2.2) and $h \varepsilon H(f)$ let

$$r(g,h:Q) = \sup_{\substack{t \,\varepsilon\, R_+ \\ x \,\varepsilon\, Q}} \{\|g(t,x) - h(t,x)\|\} \tag{5.48}$$

which we now use in the following definition for the stability of solutions of (5.46) under disturbances from the hull of $f(t,x) \varepsilon AP(C)$.

Definition 5.9: *Stability under disturbances from the hull (Sell, 1967)*

If for any $\xi > 0$ there exists a $\delta(\xi) > 0$ such that $\|y(t+\tau) - x(t;x_o,g,0)\| \leq \xi$ for $t \geq 0$ whenever $g \varepsilon H(f)$, $\|y(\tau) - x_o\| \leq \delta(\xi)$ and $r(f_\tau,g;Q) \leq \delta(\xi)$ for some $\tau \geq 0$, where $x(t;x_o,g,\tau)$ is a solution of $\dot{x} = g(t,x)$ with $x(0;x_o,g,0) = x_o$ through (τ,x_o) and $x(t;x_o,g,\tau) \varepsilon Q$ for all $t \geq \tau$. Then the solution y(t) of (5.46) for $f \varepsilon AP(C)$ is said to be *stable under disturbances from H(f)* with respect to Q.

This definition of stability for almost periodic systems is formally equivalent to the *Σ-stability* introduced by Seifert (1966). An obvious conclusion from this definition is:

Theorem 5.17:

Given that y(t) is a solution of (5.46) for $f \varepsilon AP(C)$ and is such that $\|y(t)\| \leq \beta < \alpha_1$ for all $t \geq 0$. Then if y(t) is totally stable for $t \geq 0$ it is also stable under disturbances from H(f) with respect to $Q = \{x: \|x\| \leq \gamma,\ \beta \leq \gamma \leq \alpha_1\}$, and the solution y(t) is asymptotically almost periodic in t.

A parallel set of results hold for f(t,x) periodic by which an exact equivalence exists between stability under disturbances and uniform stability. Since in the case of $f(t,x) = f(t+\omega,x)$, $\omega > 0$, total stability of (5.46) implies uniform stability and theorem **5.17** holds equally for f(t,x) periodic in t. This equivalence is not in general true for almost periodic f(t,x), although some exceptions do exist (see Kato, 1970; Yoshizawa, 1975).

5.7 Sufficient Conditions for Stability

The majority of necessary and sufficient conditions for stability of linear non-stationary homogeneous systems $\dot{x} = A(t)x$ involve the fundamental matrix X(t), which in turn implies full knowledge or computation of the solution of the systems equations. Only when the coefficient matrix A(t) is periodic, diagonal dominant or time invariant can stability conditions be directly investigated from the elements of A(t). However it is possible to generate a set of inequalities (called Wazewski's inequalities, 1958) for the sufficient conditions for stability of the linear homogeneous system

$$\dot{x} = A(t)x, \qquad A(t) \in M_n, \qquad x(t_o) = x_o \tag{5.49}$$

Theorem 5.18: *Sufficient conditions for stability (Wazewski, 1958)*

If $\lambda_{max}(t)$ and $\lambda_{min}(t)$ are the largest and smallest characteristic values of the symmetrical matrix $H(t) = A(t) + A^*(t)$, then any solution of (5.49) satisfies,

$$||x_o|| \exp\left\{\tfrac{1}{2}\int_{t_o}^{t} \lambda_{min}(s)ds\right\} \le ||x(t;x_o,t_o)|| \le ||x_o||\exp\left\{\tfrac{1}{2}\int_{t_o}^{t}\lambda_{max}(s)ds\right\} \tag{5.50}$$

Proof: The derivative of the inner products $\alpha(t) = x^*(t)x(t)$ along the solution of (5.49) is

$$\dot{\alpha} = (\dot{x})^* x + x^* \dot{x} = x^* H(t) x.$$

Then from the definitions of $\lambda_{min}(t)$ and $\lambda_{max}(t)$ of $H(t)$ (c.f. Rayleigh quotients)

$$\lambda_{min}(t) x^* x \le x^* H x \le \lambda_{max}(t) x^* x,$$

that is

$$\lambda_{min}(t) \le \dot{\alpha}\, \alpha^{-1} \le \lambda_{max}(t),$$

which on integrating gives inequality (5.50). The following sufficients conditions for stability of (5.49) are as a result of theorem **5.18.**

Corollary

The null solution to the linear system (5.49) is

(i) stable if for all $t_o \in R$,

$$\lim_{t\to\infty} \int_{t_o}^{t} \lambda_{max}(s)\, ds < N(t_o)$$

and uniformly stable if N is independent of t_o.

(ii) Unstable if

$$\lim_{t\to\infty} \int_{t_o}^{t} \lambda_{min}(s)\, ds = +\infty,$$

(iii) asymptotically stable if for all $t_o \in R$

$$\lim_{t\to\infty} \int_{t_o}^{t} \lambda_{max}(s)\, ds = -\infty,$$

and uniformly asymptotically stable if the above holds uniformly with respect to t_o.

These sufficient conditions are highly conservative and are dependent upon the particular state space representation used. The *Abel-Jacobi-Liouville lemma* (**3.2**) can be used to establish similar sufficient conditions for stability (and instability), but in this case the trace of the coefficient matrix A(t) is utilised, rather than the characteristic values of $H(t) = A(t) + A^*(t)$. Since the trace of a matrix is the sum of its characteristic values the following *instability* condition is obvious from the above corollary:

Theorem 5.19

The null solution of (5.49) is unstable if

$$\lim_{t\to\infty} \int_{t_o}^{t} \text{trace } (A(s))ds = +\infty$$

and *not* asymptotically stable if for some t_o there is a bound β such that

$$\lim_{t\to\infty} \int_{t_o}^{t} \text{trace } (A(s))ds \geq -\beta.$$

5.8 Notes and Input-Output Stability

Throughout this chapter the majority of stability results have been for linear nonautonomous systems; in almost all cases there has been a requirement to compute the system fundamental matrix, the only exceptions being in the cases of Coppel's theorem, Gershgorin's type results and for periodic systems where time invariant type results based upon system characteristic values are appropriate. Other approaches to the stability of differential equations exist, these are essentially energy methods of which Liapunov's direct approach is most notable (Venkatesh, 1977; Willems, 1970; Yoshizawa, 1966; LaSalle and Lefschetz, 1961). Of particular relevance to the study of periodic and almost periodic systems is the text of Yoshizawa (1975) which is based entirely upon Liapunov's direct method.

An alternative approach (Desoer and Vidyasagar, 1975) to Liapunov stability is based upon the input-output properties of a system rather than its internal structure as specified by differential equations. The input-output approach to stability is motivated by engineers' desire to consider systems in the frequency domain via transfer functions rather than in the time domain. The major disadvantage of the input-output technique is that it yields only the conditions for global asymptotic stability and estimation of the domain of attraction in local stability questions is not viable.

Input-output stability conditions for linear time-varying systems are readily derived from the properties of L^P Norms. Consider the forced linear time-varying system

$$\dot{x} = A(t)x + B(t)u, \tag{5.51}$$

where $A(t) \varepsilon M_n$, $B(t) \varepsilon M_m$, $u \varepsilon L_m^P[0,\infty)$, $x \varepsilon E^n$ and $x(t_o) = 0$. The solution to (5.51) can be written in terms of the state transition matrix $\Phi(t,t_o)$ of the homogeneous equation $\dot{x} = A(t)x$ as

$$x(t) = \int_{t_o}^{t} \Phi(t,t_o)\Phi(s,t_o)^{-1}B(s)u(s)ds$$

$$= \int_{-\infty}^{t} G(t,s)u(s)ds \tag{5.52}$$

where $G(t,s) \triangleq \Phi(t,t_o)\Phi(s,t_o)^{-1}B(s)$ is a $(n\times m)$ nonanticipative bounded matrix for all $t \varepsilon [t_o,\infty)$, (that is $G(t,s) = 0$ for $s > t$). The system (5.52) is said to be $L_n^P[0,\infty)$ *bounded input-output stable* when an input $u \varepsilon L_m^P[0,\infty)$ produces an output $x = Gu \varepsilon L_n^P$ or equivalently there exists a constant $0 < \gamma < \infty$ such that $||x||_p \le ||Gu||_p \le \gamma||u||_p$ whenever $u \varepsilon L_m^P$ for $1 \le p \le \infty$.

Taking norms of (5.52) and utilizing Holder's inequality for $\frac{1}{p} + \frac{1}{q} = 1$ gives

$$||x|| \le \int_{t_o}^{t} ||G(t,s)||\ ||u(s)||ds$$

$$\leq \int_{t_o}^{t} \|G(t,s)\|^{\frac{1}{p}} \|G(t,s)\|^{\frac{1}{q}} \|u(s)\| ds$$

$$\leq \left\{ \int_{t_o}^{t} \|G(t,s)\| \ \|u(s)\|^{p} \ ds \right\}^{\frac{1}{p}} \left\{ \int_{t_o}^{t} \|G(t,s)\| ds \right\}^{\frac{1}{q}} \quad (5.53)$$

Setting $\infty > \alpha \geq \int_{t_o}^{t} \|G(t,s)\| ds$ for all $t \in R_+$ then inequality (5.53) becomes, on taking the ℓ_p vector norm,

$$\|x\|_p \leq \alpha^{\frac{p}{q}} \int_{t_o}^{t} \left\{ \int_{t_o}^{t} \|G(\tau,s)\| \ \|u(s)\|^{p} ds \right\} d\tau$$

$$\leq \alpha^{\frac{p}{q}} \int_{t_o}^{t} \|u(s)\|^{p} ds \int_{t_o}^{t} \|G(\tau,s)\| \ d\tau$$

$$\leq \alpha^{(\frac{1}{p} + \frac{p}{q})} \|u\|_p , \qquad \text{for} \quad (\frac{1}{p} + \frac{1}{q}) = 1 \quad (5.54)$$

Clearly $x \in L_n^p$ and the system (5.51) is L_n^p-input/output stable. The boundedness condition on $\int_{t_o}^{t} \|G(t,s)\| ds$ for all $t \in R_+$ is a necessary and sufficient condition for L_n^{∞}. Unlike time invariant systems, linear time-varying systems can be L_n^{∞}-stable but not L_n^1-stable. For linear time invariant systems L_n^1 stability is a necessary and sufficient condition for L_n^{∞} stability, in addition $G(t,s) = G(t-s)$ and the system (5.51) is asymptotically stable in the large if and only if it is L_n^p-input/output stable (for any $1 \leq p \leq \infty$). For recent results on input/output stability for nonlinear multivariable systems see Harris and Owens (1979) and Valenca and Harris (1979).

References

Caligo, D. (1940). *Atti 2° Congresso Un.Mat.Ital.*, 177-185
Cesari, L. (1940). *Ann.Scuola Norm.Sup.Pisa.(2)* **9**, 163-186
Conley, C.C. and Miller, R.K. (1965). *J.Differential Eqns.* **1**, 333-336
Conti, R. (1955). *Riv.Mat.Unv.Parma.* **6**, 3-55
Coppel, W.A. (1965). "Stability and Asymptotic Behaviour of Differential Equations", Heath, Boston
Coppel, W.A. (1967). *Ann.Mat.Pura Appl.* **76**, 27-50
Desoer, C.A. (1970). "Notes for a Second Course on Linear Systems", Van Nostrand Reinhold,New York
Desoer, C.A. and Vidysgar, M. (1975). "Feedback Systems: Input-Output Properties", Academic Press, New York
Fink, A.M. (1974). "Almost Periodic Differential Equations", Lecture Notes in Mathematics No.377, Springer Verlag, New York
Gantmacher, F.R. (1959). "The Theory of Matrices", Vols.I,II, Chelsea, New York
Hahn, W. (1963). "Theory and Application of Liapunov's Direct Method", Prentice Hall, New Jersey
Harris, C.J. and Owens, D.H. (1979). "Multivariable Control Systems", *IEE Control and Science Record*, June 1979
Kato, J. (1970). *Tohoku Math.J.*, **22**, 254-269
LaSalle, J.P. and Lefeschetz, S. (1961). "Stability by Liapunov's Direct Method with Applications", Academic Press, New York
Lyascenko, N.Ya. (1954). *Dokl.Akad.Nank.SSSR*, **96**, 237-239
Massera, J.L. (1949). *Ann.Maths.* **50**, 705-721
Massera, J.L. (1958). *Ann.Maths.* **64**, 182-206
Massera, J.L. and Schaffer, J.J. (1958). *Ann.Maths.* **67**, 517-572
Perron, O. (1930). *Math.Zeits.* **32**, 465-473
Rapoport, I.M. (1954). "On some asymptotic methods in the theory of differential equations". *Kiev.Izdat.Akad. Nauk.Ukrain SSR*
Riesz, F. and Nagy, B.Sz. (1955). "Functional Analysis", Ungar, New York
Seifert, G. (1966). *J.Differential Eqns.*, **2**, 305-319
Seifert, G. (1968). *J.Math.Anal.Appl.*, **21**, 136-149
Strauss, A. (1969). *J.Differential Eqns.* **6**, 452-483
Valenca, J.M.E. and Harris, C.J. (1979). *Proc.IEE*, **126**, 623-627
Venkatesh, Y.V. (1977). "Energy Methods in Time-varying System Stability and Instability Analyses", LNI Physics No.68, Springer Verlag, Berlin
Wazewski, T. (1958). *Studia Mathematica* **10**, 48-59
Willems, J.L. (1970). "Stability Theory of Dynamical Systems", Nelson, London
Yoshizawa, T. (1966). "Stability Theory by Liapunov's Second Method", The Math.Soc.Japan, Tokyo
Yoshizawa, T. (1975). "Stability Theory and the Existence of Periodic Solutions and Almost Periodic Solutions", Appl.Maths. Sci. No.14, Springer Verlag, New York

Chapter 6

ASYMPTOTIC FLOQUET THEORY

6.1 Introduction

In this chapter we continue the investigation of kinematic similarity by examining the concept in the context of almost periodic matrices with particular application to the study of linear differential equations with almost periodic coefficients. It has been suggested that a generalisation of classical Floquet theory to include the almost periodic case should be of considerable use in obtaining new theorems on kinematic similarity and the stability of differential equations. Unfortunately no such generalisation is known to exist. The situation is illustrated quite simply by means of the following example:

The proposal is that for the *scalar* equation $\dot{x}(t) = a(t)x(t)$, say, with $a(t):R \rightarrow AP_1$, an analogy with the purely periodic case is sought whereby the solutions of the equation exist and assume the form

$$x(t) = p(t)\exp(bt) \tag{6.1}$$

In this representation (c.f. equations (3.83) and (4.38)) $p(t)$ possesses the characteristics of a Liapunov transformation and may even be almost periodic. Also $b \in \bar{M}_1$. We observe that $p(t)$ formally satisfies the differential equation

$$\dot{p}(t) = [a(t)-b]p(t) \tag{6.2}$$

with the implication that $a(t) \sim b$ on R. Moreover, one solution of (6.2) is

$$p(t) = \exp\left[\int_0^t (a(s)-b)ds\right] \qquad (6.3)$$

where the lower limit of integration has been taken somewhat arbitrarily as 0 for convenience. For $p(t)$ to be a generalised Liapunov transformation it must be bounded on R. However, $\int_0^t (a(s)-b)ds$ need not be bounded, even though $a(t) \in AP_1$. For if $f(t)$ and $\int_0^t f(s)ds$ are almost periodic, then $a(f,0) = 0$ is obviously necessary, otherwise the integral would contain a term $a(f,0)t$ which becomes unbounded as $|t| \to \infty$. The Fourier series of $\int_0^t f(s)ds$ is

$$a(f,0)t + \Sigma \frac{a(f,\Lambda)}{i\Lambda} \exp(i\Lambda t)$$

and an immediate question is the sufficiency of $a(f,0) = 0$ for $\int_0^t f(s)ds$ to be almost periodic. In fact it is not. To see this consider the series

$$f(t) \approx \sum_{k=1}^{\infty} \frac{1}{k^2} \exp\left(\frac{it}{k^2}\right) \qquad (6.4)$$

This series converges uniformly and so its sum $f(t)$ is almost periodic. However,

$$\int_0^t f(s)ds \approx \sum_{k=1}^{\infty} \frac{1}{i} \exp\left(\frac{it}{k^2}\right) \qquad (6.5)$$

which is not almost periodic since the coefficients violate Parseval's equation.

We shall see later that if $|\Lambda| \geq m > 0$, then $\int_0^t f(s)ds$ is almost periodic. This is the only known *simple* condition on the Fourier series which yields the almost periodicity of $\int_0^t f(s)ds$, except for the obvious condition

$$\Sigma\left|\frac{a(f,\Lambda)}{\Lambda}\right| < \infty.$$

In the remainder of this chapter we consider partial analogues of Floquet's theorem by imposing particular conditions on the almost periodic matrix F(t). The problem here is to be distinguished from the *almost constant* coefficient case in which F(t) is decomposed into two parts as follows:

$$F(t) = B + A(t)$$

It has been shown (Bellman, (1953)) that if $B \in \overline{M}_n$ has simple characteristic values λ and $\|A(t)\| \to 0$ as $t \to \infty$, then corresponding to each characteristic value there is a solution of $\dot{x}(t) = [B+A(t)]x(t)$ satisfying

$$\lim_{t\to\infty} t^{-1}\log\|x(t)\| = \operatorname{Re}\lambda \tag{6.6}$$

that is to say, we are able to evaluate the generalised characteristic exponents of F(t).

For the case in which $A(t) \in AP_n$, the requirement on A(t) can be relaxed to the extent that $\|A(t)\|$ need only be *small* (Berkey, 1976) or that some kind of *nonresonance* condition be satisfied (Coppel, 1967). Alternatively, small parameter techniques have been employed (Shtokalo, 1946, 1960; Kohn, 1976) to construct Liapunov transformations as formal power series with quasi-periodic coefficients. Other results for second order systems and systems whose structure is canonical can be found in Lyascenko (1956), Gel'man (1957, 1959, 1965), Adrianova (1962) who generalises Gel'man's earlier work and Merkis (1968) who deals with the case in which the matrix of coefficients commutes with its integral (see similar results in Chapter 4, Theorems **4.13**, **4.14**).

An essential part of the framework required to prove Coppel's theorem *(op.cit.)* is contained in the next section. We begin with the observation that new almost periodic functions can be generated from a given one by convoluting it with other functions. This result is used to prove the Coppel-Bohr lemma which is the first step in showing that the integral of an almost periodic function whose exponents are bounded *away* from zero is almost periodic.

6.2 The Coppel-Bohr Lemma and Linear Differential Equations with Almost Periodic Coefficients

The generation of almost periodic functions by convolution is especially interesting if Fourier transforms are used. Fink (1974) illustrates the idea as follows: to multiply the Fourier coefficients of $f(t) \in AP_1$ by a sequence $\{b_k\}$ is equivalent to convoluting $f(t)$ with a function whose Fourier transform is b_k at the frequencies λ_k.

Lemma 6.1: *(Fink, 1974)*

Let $f(t) \in AP_1$ and ϕ be a complex valued function such that $\hat{\phi} \in L_1(R)$ (the ^ "hat" notation means Fourier transform) and the inverse transform exists. Define

$$h(s) = \int_{-\infty}^{\infty} f(s+t)\hat{\phi}(t)dt \qquad (6.7)$$

Then $h(t) \in AP_1$ and

$$h(t) \approx \Sigma\ a(h,\lambda)\phi(\lambda)\exp(i\lambda t) \qquad (6.8)$$

In addition

$$T(\eta,h(t)) \supset T(\frac{\eta}{\|\phi\|_1}, f(t)) \qquad (6.9)$$

where $\|\cdot\|_1$ is the L_1 norm.

Proof (see Fink, (op.cit.))

The Coppel-Bohr lemma is really a result about trigonometric polynomials. It is used as the starting point for the Approximation Theorem (Theorem **2.13**) which extends the result.

Lemma 6.2: *(Coppel, 1967)*

Let $f(t)$ be a trigonometric polynomial

$$f(t) = \sum_{k=1}^{N} a_k \exp(i\lambda_k t) \qquad (6.10)$$

with $|\lambda_k| \geq m$ for all k. Let

$$g(t) = \sum_{k=1}^{N} \frac{a_k}{i\lambda_k} \exp(i\lambda_k t). \tag{6.11}$$

Then there exists a constant d independent of f, g and m such tha

$$\|g\| \le dm^{-1}\|f\|. \tag{6.12}$$

Proof is taken from Fink *(op.cit.)*.

Noting that g(t) is the unique integral of f(t) with mean valu zero, the idea of Lemma **6.1** is employed. Consider first of all the case when $m = 1$ and define

$$\phi(t) = \begin{cases} it & \text{for} \quad |t| \le 1 \\ (-it)^{-1} & \text{for} \quad |t| \ge 1 \end{cases} \tag{6.13}$$

For $s \neq 0$ we can estimate the Fourier transform of $\phi(t)$ as follows:

$$2\pi\hat{\phi}(s) = \int_{-\infty}^{\infty} \phi(t)\exp(-ist)dt \tag{6.14}$$

Integrating by parts once yields

$$2\pi\hat{\phi}(s) = \frac{1}{is}\int_{-\infty}^{\infty} \dot{\phi}(t)\exp(-ist)dt \tag{6.15}$$

and once again

$$= \frac{1}{s}\int_{-1}^{1} \exp(-ist)dt - \frac{i\exp(-ist)}{(st)^2}\Bigg|_{+1}^{-1} - \frac{2i}{s^2}\int_{|t|\ge 1} \frac{\exp(-ist)}{t^3}dt \tag{6.16}$$

$$= \frac{2i}{s^2}[\exp(-is)-\exp(is)] - \frac{2i}{s^2}\int_{|t|\ge 1} \frac{\exp(-ist)}{t^3}dt \tag{6.17}$$

If $|t| \ge 1$ then $t^3 \ge \frac{1}{2}(1+t^2)$. Hence

$$2\pi\hat{\phi}(s) \le 2s^{-2} + 2s^{-2}\int_{|t|\ge 1} \frac{dt}{\frac{1}{2}(1+t^2)} = cs^{-2} \tag{6.18}$$

where c is some constant, and

$$2\pi\hat{\phi}(s) \leq 2c(1+s^2)^{-1} \qquad \text{if} \quad |s| \geq 1. \tag{6.19}$$

For $|s| \leq 1$ and $s \neq 0$ we have

$$|2\pi\hat{\phi}(s)| = \left| \int_{-1}^{1} it\exp(-ist)dt + i\int_{|t|\geq 1} \frac{\exp(-ist)}{t} dt \right| \tag{6.20}$$

$$\leq 2 + \left| \int_{1}^{\infty} \frac{\sin st}{t} dt \right| = 2 + \pi$$

$$\leq 2(2+\pi)(1+s^2)^{-1} \tag{6.21}$$

Since $\hat{\phi}(s)$ is invertible, we have

$$g(t) = -\sum_{k=1}^{N} a_k \phi(\lambda_k)\exp(i\lambda_k t)$$

$$= -\sum_{k=1}^{N} a_k \int_{-\infty}^{\infty} \hat{\phi}(s)\exp(i\lambda_k s)ds \exp(i\lambda_k t)$$

$$= -\int_{-\infty}^{\infty} \left[\sum_{k=1}^{N} a_k \exp(i\lambda_k(s+t)) \right] \hat{\phi}(s)ds \tag{6.22}$$

$$= -\int_{-\infty}^{\infty} f(s+t)\hat{\phi}(s)ds \tag{6.23}$$

whence

$$\|g\| \leq \|\hat{\phi}\|_1 \|f\|. \tag{6.24}$$

Note that $\|\hat{\phi}\|_1$ does not depend on f, g or Λ but is an absolute constant. For the general case, suppose $|\lambda_k| \geq m$ and consider

$$h(t) = f(\tfrac{t}{m}) \tag{6.25}$$

$$w(t) = mg(\tfrac{t}{m}) \tag{6.26}$$

Then

$$h(t) = \sum_{k=1}^{N} a_k \exp(i \frac{\lambda_k}{m} t) \tag{6.27}$$

and

$$w(t) = \sum_{k=1}^{N} \frac{a_k}{i(\frac{\lambda_k}{m})} \exp(i \frac{\lambda_k}{m} t) \tag{6.28}$$

Clearly the exponents $\mu_k = \frac{\lambda_k}{m}$ satisfy $|\mu_k| \geq 1$. Using the above result we then have

$$\|w\| \leq \|\hat{\phi}\|_1 \|h\| \tag{6.29}$$

so that

$$\|g\| = m^{-1}\|w\| \leq m^{-1}\|\hat{\phi}\|_1 \|f\| \tag{6.30}$$

Replacing $\|\hat{\phi}\|_1$ by d gives the required result.

Lemma 6.3: *(Levitan, 1953)*

Let $f(t) \in AP_1$ such that

$$f(t) \approx \sum a_k \exp(i\lambda_k t)$$

with $|\lambda_k| \geq m > 0$ for all k. Then

$$\int_0^t f(s)ds \in AP_1.$$

If

$$g(t) = \int_0^t f(s)ds \qquad \text{with} \quad a(g,0) = 0,$$

then

$$\|g\| \leq dm^{-1}\|f\| \tag{6.31}$$

where d is an absolute constant.

Proof: The lemma has already been proved for trigonometric polynomials. The Approximation Theorem extends the result as required (see Fink *(op.cit.)* and Coppel *(op.cit.)*).

We are now in a position to say something about solutions of the inhomogeneous linear equation

$$\dot{x}(t) = Bx(t) + f(t) \tag{6.32}$$

where $B \in \overline{M}_n$, $f(t) = \mathrm{col}(f_1(t), f_2(t), \ldots, f_n(t))$, $f_k: R \to AP_1$, $x: R \to \mathbb{E}^n$. Scalar equations will be considered initially and then the vector equations will be built from short sequences of the scalar equations. The aim is to show that the vector equation (6.32) can have almost periodic solutions, even if the corresponding homogeneous equation has almost periodic solutions, provided that the Fourier exponents of these solutions are not arbitrarily close to the Fourier exponents of any of the f_k. This is clearly a nonresonance condition.

Lemma 6.4: *(Coppel, 1967)*

Suppose $b = i\beta$ for some real β such that $|\beta - \lambda_k| \geq m > 0$ for all $\lambda_k \in \Lambda_f$. Then there exists a unique almost periodic solution $x(t)$ to

$$\dot{x}(t) = bx(t) + f(t) \tag{6.33}$$

such that $\Lambda_x = \Lambda_f$. Moreover,

$$\|x\| \leq dm^{-1}\|f\|$$

where d is the numerical constant of Lemma **6.3**.

Proof: The change of variables

$$y(t) = \exp(-i\beta t)x(t) \tag{6.34}$$

transforms equation (6.33) into

$$\dot{y}(t) = \exp(-i\beta t)f(t) \equiv g(t) \tag{6.35}$$

At this point we observe that

$$\|g\| = \|f\|, \qquad \|x\| = \|y\| \tag{6.36}$$

with

$$\Lambda_y = \Lambda_x - \beta \tag{6.37}$$

and similarly

$$\Lambda_g = \Lambda_f - \beta \tag{6.38}$$

If the nonresonance condition is satisfied then the exponents of g(t) are bounded away from zero by virtue of (6.38). Lemma **6.3** asserts that y(t) is the unique integral of g(t) with $M_t(y) = 0$, $y:R \to AP_1$ and

$$\|y\| \leq dm^{-1}\|g\|.$$

Thus the sets of exponents of y(t) and g(t) are the same, so by reversing the change of variables (6.34) we obtain the desired result.

The following lemma deals with the case in which b is complex:

Lemma 6.5

Suppose that $Re(b) \neq 0$ and $f:R \to AP_1$. Then there exists a unique almost-periodic solution x(t) to (6.33) such that $\Lambda_x = \Lambda_f$. Moreover,

$$\|x\| \leq |Re(b)|^{-1}\|f\| \tag{6.35}$$

Proof: We seek bounded solutions to (6.33) for $Re(b) \neq 0$. Two possibilities arise. The first is $Re(b) > 0$, in which case

$$x(t) = -\int_t^{\infty} \exp(b(t-s))f(s)ds \tag{6.36}$$

is the required solution with

$$\|x\| \leq Re(b)^{-1}\|f\| \tag{6.37}$$

The other possibility is that $Re(b) < 0$ and then

$$x(t) = \int_{-\infty}^{t} \exp(b(t-s))f(s)ds \tag{6.38}$$

with

$$\|x\| \leq -Re(b)^{-1}\|f\| \tag{6.39}$$

The change of variables $u = s-t$ in (6.36) gives

$$x(t) = -\int_0^{\infty} \exp(-bu)f(u+t)dt \tag{6.40}$$

whence

$$|x(t+\tau)-x(t)| \leq \mathrm{Re}(b)^{-1}\|f(t+\tau)-f(t)\| \tag{6.41}$$

Clearly x(t) is almost periodic with

$$T(\eta,x(t)) \supset T(\eta \mathrm{Re}(b),f(t)) \tag{6.42}$$

To show that $\Lambda_x = \Lambda_f$ we consider the Fourier series

$$x(t) \approx a_o + \sum_{k=1}^{\infty} a_k \exp(i\lambda_k t) \tag{6.43}$$

$$\dot{x}(t) \approx \sum_{k=1}^{\infty} i\lambda_k a_k \exp(i\lambda_k t) \tag{6.44}$$

The differential equation then gives

$$\begin{aligned} i\lambda_k a_k - ba_k &= a(f,\lambda_k) \qquad \text{for} \quad k \geq 1 \\ - ba_o &= a(f,0) \qquad \text{for} \quad k = 0 \end{aligned} \tag{6.45}$$

Since $i\lambda_k - b \neq 0$ for all k, it follows that a necessary and sufficient condition for $a_k \neq 0$ is that $a(f,\lambda_k) \neq 0$, hence $\Lambda_x = \Lambda_f$.

The above can be extended to equations where x and f have values in R^n or E^n and $b \equiv B \in \overline{M}_n$. A suitable change of variables can be imposed to triangularise B and then the scalar result applied successively. This was first proved by Bohr and Neugebauer (1926).

Theorem 6.1

Suppose that $|\beta - i\lambda_k| \geq m > 0$ for all characteristic values β of B and exponents $\lambda_k \in \Lambda_f$. Then there exists a unique almost periodic solution x(t) to (6.32) with $\Lambda_x = \Lambda_f$.

Furthermore, there exists a polynomial Γ of degree less than or equal to n with no constant term, depending only on the matrix B and an absolute constant d, so that

$$\|x\| \leq \Gamma(dm^{-1})\|f\| \tag{6.46}$$

Proof: Following Fink *(op.cit.)*, we first demonstrate existence. Noting that for any matrix $B \in \overline{M}_n$ there exists a $P \in \overline{M}_n$ such that $PBP^{-1} = T$ is a lower triangular matrix, the change of variables $y = Px$ in equation (6.32) gives:

$$\dot{y}(t) = Ty(t) + g(t) \tag{6.47}$$

where $g = Pf$. It is clear that $\Lambda_y \subset \Lambda_x$ by the change of variables and the reverse is also true. Hence $\Lambda_y = \Lambda_x$. Similarly

$$\|x\| \, \|P^{-1}\|^{-1} \le \|y\| \le \|P\| \, \|x\|$$

The first equation in (6.47) is

$$\dot{y}_1(t) = t_{11}y_1(t) + g_1(t) \tag{6.48}$$

which is of the form of the scalar equation discussed in Lemma **6.4**. Because of the form of T, t_{11} is in fact a characteristic value of B. Subject to the hypothesis of the theorem we apply Lemma **6.4** and deduce that there exists a unique almost periodic solution of (6.48) with $\Lambda_{y_1} = \Lambda_{g_1}$ and

$$\|y_1\| \le dm^{-1}\|g_1\| \tag{6.49}$$

The second scalar equation in (6.32) is

$$\dot{y}_2(t) = t_{22}y_2(t) + (g_2(t)+t_{21}y_1(t))$$

$$= t_{22}y_2(t) + h_2(t) \tag{6.50}$$

where

$$h_2(t) = g_2(t) + t_{21}y_1(t)$$

is almost periodic with $\Lambda_{h_2} \subset \Lambda_f$. With the above hypothesis we get a unique almost periodic $y_2(t)$ whose exponents $\Lambda_{y_2} = \Lambda_{h_2} \subset \Lambda_f$ which satisfies the estimate

$$\begin{aligned}\|y_2\| &\le dm^{-1}\|h_2\| \\ &\le dm^{-1}(\|g_2\|+|t_{21}|\cdot\|y_1\|) \\ &\le dm^{-1}(\|g_2\|+|t_{21}|dm^{-1}\|g_1\|) \\ &\le dm^{-1}(\|g\|+|t_{21}|dm^{-1}\|g\|) \\ &= dm^{-1}\|g\|(1+|t_{21}|dm^{-1})\end{aligned} \tag{6.51}$$

Similar estimates can be found for the remaining elements $y_3 \ldots$ $\ldots y_n$ of y and finally we obtain a solution y(t) with $\Lambda_y \subset \Lambda_f$ which satisfies the estimate

$$\|y\| \leq \|g\|\Gamma(dm^{-1}) \tag{6.52}$$

where $\Gamma(\cdot)$ is a polynomial of degree n with no constant term. Evidently the coefficients of Γ depend on the matrix T and therefore the matrix B. Using the relationship between the norms of x and y, the estimate (6.52) can be transferred back to the original equation.

The uniqueness of y(t) (and hence of x(t)) remains to be shown. If there are two solutions $x_\alpha(t)$ and $x_\beta(t)$ of (6.32) with

$$\Lambda_{x_\alpha} \subset \Lambda_f, \qquad \Lambda_{x_\beta} \subset \Lambda_f$$

then $x_\alpha(t) - x_\beta(t)$ is also an almost periodic solution whose exponents are contained in Λ_f. The Fourier series of $x_\alpha(t)$ and $x_\beta(t)$ are the following

$$\begin{aligned} x_\alpha(t) &\approx \sum_{k=1}^{\infty} a_k \exp(i\lambda_k t) \\ x_\beta(t) &\approx \sum_{k=1}^{\infty} b_k \exp(i\lambda_k t), \qquad \lambda_k \in \Lambda_f \end{aligned} \tag{6.53}$$

whence

$$\begin{aligned} x_\alpha(t) - x_\beta(t) &\approx \sum_{k=1}^{\infty} (a_k - b_k)\exp(i\lambda_k t) \\ \dot{x}_\alpha(t) - \dot{x}_\beta(t) &\approx \sum_{k=1}^{\infty} i\lambda_k(a_k - b_k)\exp(i\lambda_k t) \end{aligned} \tag{6.54}$$

Since

$$\dot{x}_\alpha(t) - \dot{x}_\beta(t) = B(x_\alpha(t) - x_\beta(t))$$

it follows that

$$i\lambda_k(a_k - b_k) = B(a_k - b_k).$$

According to the hypothesis of the theorem $i\lambda_k$ is not a characteristic value of B so $a_k - b_k = 0$ for all $\lambda_k \in \Lambda_f$. By the uniqueness theorem $x_\alpha(t) = x_\beta(t)$. Linearity follows from uniqueness.

The final step before discussing Coppel's theorm *(op.cit.)* is to consider the matrix differential equation

$$\dot{U}(t) = BU(t) - U(t)B + C(t) \tag{6.55}$$

with $U(t) \in M_n$, $C(t) \in AP_n$ and $B \in \overline{M}_n$. The equation can be rewritten so that the right hand side of the homogeneous part is a linear mapping with characteristic values $(\beta_j - \beta_\ell)$, $j,\ell=1,\ldots,n$. Obviously 0 belongs to the set $\{\beta_j - \beta_\ell\}$. This implies that in order to avoid resonance, the exponents of C(t) must be restricted in some way and in fact they are required to be bounded away from zero by a distance m. Theorem **6.1** can now be applied to give a unique almost periodic solution of (6.55) with $\Lambda_u = \Lambda_c$ which satisfies the estimate

$$\|U\| \leq \|C\|\Gamma(dm^{-1}) \tag{6.56}$$

where $\|\cdot\|$ is the appropriate matrix norm. The case in which B is a diagonal matrix whose characteristic values have distinct real parts is also very interesting. Here we can apply Lemma **6.5** directly and obtain the estimate

$$\|U\| \leq \frac{1}{\min\limits_{\substack{1\leq j,\ell\leq n \\ j\neq\ell}} |\mathrm{Re}\beta_\ell - \mathrm{Re}\beta_j|} \|C\| \tag{6.57}$$

6.3 Coppel's Theorem

For the almost periodic case, one of the most significant partial analogues to Floquet's theorm is Coppel's theorem *(op.cit.)*.

Theorem 6.2: *(Coppel, 1967)*

Consider the linear homogeneous differential equation

$$\dot{x}(t) = [B+A(t)]x(t) \tag{6.58}$$

with $x:R \to E^n$, $B \in \overline{M}_n$ and $A:R \to AP_n$. Let the exponents of $A(t)$ be denoted λ_k and let $S(A)$ denote the set of all numbers which are linear combinations of λ_k with non-negative integral coefficients, at least one of which is positive. Suppose that

$$\text{dist}[\beta_j-\beta_\ell,\ iS(A)] \quad > \quad 0 \tag{6.59}$$

for all $j,\ell = 1,\ldots,n$, where $\beta_\ell (\ell=1,\ldots,n)$ are the characteristic values of B. Then a fundamental matrix $X(t)$ of (6.58) is of the form

$$X(t) \quad = \quad P(t)\exp(Bt) \tag{6.60}$$

where $P(t) = I+Q(t)$, $Q:R \to AP_n$ and $\Lambda_Q \subset S(A)$.

Proof: The linear transformation (6.60) applied to equation (6.58) yields the following equation for $P(t)$:

$$\dot{P}(t) \quad = \quad BP(t) - P(t)B + A(t)P(t) \tag{6.61}$$

We have already examined this kind of equation. In fact, for $C:R \to AP_n$ it is known that

$$\dot{U}(t) \quad = \quad BU(t) - U(t)B + C(t) \tag{6.62}$$

admits an estimate

$$\|U\| \quad \leq \quad \|C\|\Gamma(dm^{-1}) \tag{6.63}$$

This will be used repeatedly. Assume that $\Lambda_C \subset S(A)$. Assume also that Λ_A are all one sign. Starting from $U_o(t) = I$ we define a sequence $\{U_k(t)\}$ inductively by taking $U_k(t)$ to be the unique almost periodic solution of (6.62) with

$$C(t) \quad = \quad A(t)U_{k-1}(t) \tag{6.64}$$

so that

$$\Lambda_{U_k} \quad = \quad \Lambda_{AU_{k-1}} \tag{6.65}$$

with

$$\Lambda_{AU_{k-1}} \quad \subset \quad \Lambda_A + \Lambda_{U_{k-1}} \tag{6.66}$$

Therefore $\Lambda_{U_k} \subset S(A)$, this being true for all k. The hypothesis that $\beta_j - \beta_j = 0$ is a positive distance from iS(A) implies that if $\lambda \in S(A)$ then $|\lambda| \geq m$. Thus any exponent of $U_k(t)$, μ say, satisfies the inequality $|\mu| \geq km$. Clearly the distance m_k between $\{\beta_j - \beta_\ell\}$ and $i\Lambda_{U_k}$ increases with k so that for sufficiently large k we obtain the estimate

$$\|U_k\| \leq \Gamma(dm_k^{-1})\|A\|\cdot\|U_{k-1}\| \tag{6.67}$$

Since $\Gamma(\cdot)$ is a polynomial with no constant term and m_k grows like k, an N may be chosen so large that for $k > N$

$$\Gamma(dm_k^{-1})\|A\| < \frac{\rho}{k} \tag{6.68}$$

where $\rho = 2\|A\|\frac{d}{m}$. The inequality

$$\|U_k\| \leq \frac{\rho}{k}\|U_{k-1}\|$$

implies that $\Sigma\|U_k\|$ is majorised by $\Sigma\frac{\rho^k}{k!}$. Therefore the series

$$P(t) = \sum_{k=0}^{\infty} U_k(t) = I + Q(t) \tag{6.69}$$

converges uniformally on R and is almost periodic with $\Lambda_P \subset S(A)$. Note that Q(t) is almost periodic and $\Lambda_Q \subset S(A)$. It is easily seen that P(t) is a solution of equation (6.61). Thus $X(t) = P(t)\exp[Bt]$ is a solution of equation (6.58).

Similarly we can find an almost periodic solution

$$P_1(t) = I + Q_1(t)$$

to

$$\dot{P}(t) = BP(t) - P(t)B - P(t)A(t) \tag{6.70}$$

with $\Lambda_{Q_1} \subset S(A)$. It follows that $P_1(t)P(t)$ is a solution of the equation

$$\dot{U}(t) = BU(t) - U(t)B \tag{6.71}$$

Therefore

$$P_1(t)P(t) - I = Q_1(t) + Q(t) + Q_1(t)Q(t)$$

is a solution of the same equation and its exponents are contained in S(A). Hence it must be the zero solution. Thus $P_1(t)P(t) = I$ and the matrix $I+Q(t)$ has an inverse of the same form. Therefore $P(t)$ is a fundamental matrix solution of (6.61).

This theorem can be interpreted immediately as a result on kinematic similarity. Also the analogy with classical Floquet theory is clearly demonstrated.

Theorem 6.3

Suppose that $A:R \to AP_n$ and $B \in \overline{M}_n$. Let the exponents of $A(t)$ be denoted λ_k and let $S(A)$ denote the set of all numbers which are linear combinations of λ_k with non-negative integral coefficients. Then for $[B+A(t)] \sim B$ it is sufficient that

$$\operatorname{dist}[\beta_j-\beta_\ell,\ iS(A)] > 0$$

for all $j,\ell = 1,2,\ldots,n$, where $\beta_\ell (\ell=1,\ldots,n)$ are characteristic values of B.

Proof: follows immediately from Theorem 6.2.

Berkey (1976) considered the same equation as Coppel but with different assumptions. He proved that each independent scalar solution of the equation approaches the form

$$x_k(t) \to q_k(t)\exp[(\beta_k+M_k)t]$$

as $t \to \infty$ with q_k almost periodic and M_k the mean value of a certain almost periodic function.

Theorem 6.4: *(Berkey, 1976)*

Suppose $A:R \to AP_n$ and $B = \operatorname{diag}(\beta_1,\ldots,\beta_n) \in \overline{M}_n$ with $\operatorname{Re}(\beta_j-\beta_\ell) \neq 0$ for $j \neq \ell$. Then for $\|A\|$ sufficiently small, equation (6.58) possesses n independent solutions $x_k(t)$ of the form

$$x_k(t) = (p_k(t)+e_k)\exp\left[\beta_k t + \int_0^t v_k(s)ds\right] \tag{6.72}$$

where $e_k = \operatorname{col}(\delta_{1k},\delta_{2k},\ldots,\delta_{nk})$, $v_k(t)$ is a scalar almost periodic function and

$$p_k(t) = \operatorname{col}(p_{k,1}(t),\ldots,p_{k,k-1}(t),0,p_{k,k+1}(t),\ldots,p_{k,n}(t)) \tag{6.73}$$

where $p_{k,\ell}(t)$ and $\dot{p}_{k,\ell}(t)$ are almost periodic for $1 \le \ell \le n$.
Proof: (Berkey, op.cit.)

It is clear that the characteristic exponents of $[B+A(t)]$ admit the estimate

$$\lim_{t\to\infty} t^{-1}\log\|x_k(t)\| = \mathrm{Re}\beta_k + M_k \tag{6.74}$$

where

$$M_k = \lim_{t\to\infty} t^{-1}\int_0^t v(s)ds.$$

The requirement that $\|A(t)\| \to 0$ as $t \to \infty$ (the almost constant coefficient case) would force $M_k = 0$, $k = 1,\ldots,n$.

6.4 Almost Periodic Matrices Containing a Parameter

We consider the almost periodic matrix $A(t,\delta)$ which is analytic in δ:

$$A(t,\delta) = A_o + \sum_{k=1}^{\infty} A_k(t)\delta^k \tag{6.75}$$

with $A_o \in \overline{M}_n$, $A_k{:}R \to AP_n$ for $k = 1,2,\ldots$ and δ is a small real or complex parameter. It is assumed that $A(t,\delta)$ is almost periodic in t uniformly with respect to δ in a domain E^* and analytic in δ in E^*. The power series $\Sigma\|A_k\|\delta^k$ is assumed to converge for $|\delta| \le \rho$ and in particular for $|\delta| > 0$.

Theorem 6.5

Suppose that the matrix $A(t,\delta)$ is almost periodic in t uniformly with respect to δ in a domain E and analytic in δ in E, and can be represented by the power series

$$A(t,\delta) = A_o + \sum_{k=1}^{\infty} A_k(t)\delta^k$$

which converges for $|\delta| < \rho$. Assuming that the matrix $A_o \in \overline{M}_n$ is diagonal and possesses characteristic values with distinct real parts and $A_k{:}R \to AP_n$ for all $k = 1,2,\ldots,$ then

$$A(t,\delta) \sim B(t,\delta) \tag{6.76}$$

where

$$B(t,\delta) = B_0 + \sum_{k=1}^{\infty} B_k(t)\delta^k \tag{6.77}$$

for $|\varepsilon| < \bar{\rho} \leq \rho$, $A_0 = B_0 \in \overline{M}_n$ and $B_k : R \to AP_n$ are diagonal matrices. Furthermore

$$\underset{k}{\cup} \Lambda_{B_k} \subset S(\underset{k}{\cup}\Lambda_{A_k}) \tag{6.78}$$

The proof of the theorem will be preceded by two lemmas.

Lemma 6.6: *(Gel'man, 1959; Blinov, 1965)*

Suppose that the power series

$$\sum_{k=1}^{\infty} a_k \xi^k \tag{6.79}$$

is convergent, where $a_k \in R_+$ and $\xi \in R$. If the equation

$$\sum_{k=1}^{\infty} a_k \xi^k = 1 \tag{6.80}$$

possesses at least one root, then there exists a unique positive root $\bar{\xi}$ which is the least in absolute value of all the roots.

Lemma 6.7: *(Gel'man, 1959; Blinov, 1965)*

The function of the real variable ξ

$$p(\xi) = a(\xi)^{-1}[1-a(\xi)-[1-2a(\xi)]^{\frac{1}{2}}] \tag{6.81}$$

where $a(\xi)$ is a power series of the form (6.79), is single valued for all $|\xi| < \bar{\rho} \leq \rho$. Here ρ is the radius of convergence of (6.79). Furthermore, if the equation

$$2a(\xi) = 1 \tag{6.82}$$

has at least one root, then $\bar{\rho}$ is the smallest positive root of this equation otherwise $\bar{\rho} = \rho$.

Proof of Theorem ***6.5***

It is necessary and sufficient to show that there exists a Liapunov transformation which assures the required similarity. To complete the analogy with classical Floquet theory we suppose that $P(t,\delta)$ is the required transformation which admits a formal power series expansion as follows:

$$P(t,\delta) = I + \sum_{k=1}^{\infty} P_k(t)\delta^k \tag{6.83}$$

with $P_k : R \to AP_n$ possessing diagonal entries equal to zero. The aim is to show that the series (6.83) is convergent and to estimate its radius of convergence.

The equation

$$B(t,\delta) = -P(t,\delta)^{-1}[\dot{P}(t,\delta) - A(t,\delta)P(t,\delta)]$$

defines the supposed similarity, and may be written as

$$\dot{P}(t,\delta) = A(t,\delta)P(t,\delta) - P(t,\delta)B(t,\delta) \tag{6.84}$$

We solve this equation; the solution will be almost periodic in t uniformly in δ, analytic in δ in some domain and have an inverse.

Substituting (6.75), (6.77) and (6.83) into (6.84) yields the identity

$$\sum_{k=1}^{\infty} \dot{P}_k(t)\delta^k = \left[A_o + \sum_{k=1}^{\infty} A_k(t)\delta^k\right]\left[I + \sum_{k=1}^{\infty} P_k(t)\delta^k\right] - \left[I + \sum_{k=1}^{\infty} P_k(t)\delta^k\right]\left[B_o + \sum_{k=1}^{\infty} B_k(t)\delta^k\right] \tag{6.85}$$

and by equating like powers of δ we obtain

$$A_o = B_o \tag{6.86}$$

and a sequence of differential equations:

$$\dot{P}_1(t) = A_oP_1(t) - P_1(t)A_o + A_1(t) - B_1(t)$$

$$\dot{P}_2(t) = A_oP_2(t) - P_2(t)A_o + A_2(t) + A_1(t)P_1(t) - B_2(t) - P_1(t)B_1(t)$$

$$\dot{P}_3(t) = A_oP_3(t) - P_3(t)A_o + A_3(t) + A_2(t)P_1(t) + A_1(t)P_2(t) - B_3(t) - P_1(t)B_2(t) - P_2(t)B_1(t)$$

$$\begin{array}{ccc} \cdot & & \cdot \\ \cdot & & \cdot \end{array}$$

$$\begin{aligned} \dot{P}_k(t) = {} & A_o P_k(t) - P_k(t)A_o + A_k(t) + A_{k-1}(t)P_1(t) + \ldots \\ & + A_1(t)P_{k-1}(t) \\ & - B_k(t) - P_1(t)B_{k-1}(t) - P_2(t)B_{k-2}(t) - \ldots \\ & - P_{k-2}(t)B_2(t) - P_{k-1}(t)B_1(t) \end{aligned} \tag{6.87}$$

The choice

$$B_1(t) = \text{diag}[A_1(t)]$$

$$B_2(t) = \text{diag}[A_2(t) + A_1(t)P_1(t)]$$

$$B_3(t) = \text{diag}[A_3(t) + A_2(t)P_1(t) + A_1(t)P_2(t)]$$

$$\begin{array}{ccc} \cdot & & \cdot \\ \cdot & & \cdot \\ \cdot & & \cdot \end{array}$$

$$B_k(t) = \text{diag}[A_k(t) + A_{k-1}(t)P_1(t) + \ldots + A_1(t)P_{k-1}(t)] \tag{6.88}$$

ensures that the right hand sides of equations (6.87) have diagonal entries equal to zero. This implies that each of the equations (6.87) is essentially decoupled. Therefore the initial hypothesis concerning A_o guarantees, through Lemma **6.5**, the existence of unique almost periodic matrix solutions to (6.87). In fact,

$$\|P_1\| \leq h\|A_1\|$$

with

$$\Lambda_{P_1} = \Lambda_{A_1}$$

$$\|P_2\| \leq h[\|A_2\| + 2\|A_1\|\cdot\|P_1\|]$$

with

$$\Lambda_{P_2} \subset \Lambda_{A_2} \cup [\Lambda_{A_1} + \Lambda_{P_1}]$$

$$\|P_3\| \le h[\|A_3\|+2(\|A_1\|\cdot\|P_2\|+\|A_2\|\cdot\|P_1\|)+\|A_1\|\cdot\|P_1\|\cdot\|P_1\|]$$

with

$$\Lambda_{P_3} \subset \Lambda_{A_3} \cup [\Lambda_{A_1}+\Lambda_{P_2}] \cup [\Lambda_{A_2}+\Lambda_{P_1}] \ \ldots \ \cup [\Lambda_{A_1}+\Lambda_{P_1}+\Lambda_{P_1}]$$

and since the operation + is distributive through unions

$$\Lambda_{P_3} \subset \Lambda_{A_3} \cup [\Lambda_{A_2}+\Lambda_{A_1}] \cup [\Lambda_{A_1}+\Lambda_{A_1}+\Lambda_{A_1}]$$

$$\vdots \qquad \vdots \qquad \vdots$$

$$\begin{aligned}\|P_k\| \le\ & h[\|A_k\|+2(\|A_1\|\cdot\|P_{k-1}\|+ \ldots +\|A_{k-1}\|\cdot\|P_1\|) \\ & + \|A_1\|(\|P_1\|\cdot\|P_{k-2}\|+ \ldots +\|P_{k-2}\|\cdot\|P_1\|)+ \ldots \\ & + \|A_{k-3}\|(\|P_1\|\cdot\|P_2\|+ \ldots +\|P_2\|\cdot\|P_1\|)+ \ldots \\ & + \|A_{k-2}\|(\|P_1\|\cdot\|P_1\|)]\end{aligned} \tag{6.89}$$

with

$$\bigcup_k \Lambda_{P_k} \subset S(\bigcup_k \Lambda_{A_k}) \tag{6.90}$$

The real constant h is given by

$$h = \frac{1}{\min\limits_{\substack{1\le j,\ell\le n \\ j\ne\ell}} |\mathrm{Re}\alpha_\ell - \mathrm{Re}\alpha_j|}$$

where $\alpha_j, j = 1,2,\ldots,n$ are characteristic values of A_o.

Setting $a_k = \|A_k\|$ and $b_k = 2a_k$, the estimates (6.89) can be rewritten as

$$\|P_1\| \le ha_1$$

$$\|P_2\| \le h[a_1+b_1\|P_1\|]$$

$$\|P_3\| \le h[a_3+(b_1\|P_2\|+b_2\|P_1\|)+a_1(\|P_1\|\cdot\|P_1\|)]$$

$$\vdots \qquad \vdots \qquad \vdots$$

$$\begin{aligned}\|P_k\| \leq\ & h[a_k+(b_1\|P_{k-1}\|+\ldots+b_{k-1}\|P_1\|)+a_1(\|P_1\|\cdot\|P_{k-2}\| \\ & + \ldots + \|P_{k-2}\|\cdot\|P_1\|)+ \ldots +a_2(\|P_1\|\cdot\|P_{k-3}\| \\ & + \ldots + \|P_{k-3}\|\cdot\|P_1\|)+ \ldots +a_{k-3}(\|P_1\|\cdot\|P_2\| \\ & + \ldots + \|P_2\|\cdot\|P_1\|) + \ldots +a_{k-2}(\|P_1\|\cdot\|P_1\|)]\end{aligned} \tag{6.92}$$

Similarly set $\|P_k\| = p_k$; then (6.92) becomes (taking the equality)

$$p_1 = ha_1$$

$$p_2 = h[a_2+b_1p_1]$$

$$p_3 = h[a_3+(b_1p_2+b_2p_1)+a_1(p_1p_1)]$$

$$\vdots \qquad \vdots \qquad \vdots$$

$$\begin{aligned}p_k = h[&a_k+(b_1p_{k-1}+ \ldots +b_{k-1}p_1)+a_1(p_1p_{k-2}+ \ldots \\ & +p_{k-2}p_1)+ \ldots +a_{k-3}(p_1p_2+p_2p_1)+a_{k-2}(p_1p_1)]\end{aligned} \tag{6.93}$$

Now we define $a(\delta) = 2h \sum_{k=1}^{\infty} \|A_k\|\delta^k$. The series $\sum_{k=1}^{\infty} \|P_k\|\cdot|\delta^k|$ which consists of norms of the terms of the series (6.83) is majorised by the series $p(\xi) = \sum_{k=1}^{\infty} p_k\xi^k$ in which the p_k satisfy (6.93) and $\xi = |\delta|$, whose generating function is (6.81). According to Lemma **6.7** this function is single valued for $|\xi| < \bar{\rho}$ where $\bar{\rho}$ is the unique positive root of $2a(\xi) = 1$ if it has any, otherwise $\bar{\rho} = \rho$. This defines the radius of convergence of $p(\xi)$ which in turn implies the convergence of (6.83). We conclude that there exists an almost periodic solution to equation

(6.84) for $|\delta| < \bar{\rho} \leq \rho$ and it remains to be shown that this solution is fundamental and has an almost periodic inverse.

Recall that a solution is fundamental if and only if

$$\det[I + \sum_{k=1}^{\infty} P_k(t)\delta^k] \neq 0 \qquad \text{for all } t \tag{6.94}$$

Since

$$\|\sum_{k=1}^{\infty} P_k(t)\delta^k\| \leq p(\xi) \quad \text{and} \quad p(\xi) < 1$$

if $|\xi| < \rho$, then the condition that $\det(P(t,\delta)) \neq 0$ is clearly fulfilled. That the matrix $P(t,\delta)^{-1}$ is almost periodic has been proved by Fink (1971) who observes that it is sufficient to show that $\det[P(t,\delta)]^{-1}$ is almost periodic. Each component of adjoint $P(t,\delta)$ is a polynomial in the components of $P(t,\delta)$ and for that reason is almost periodic. For the same reason $\det[P(t,\delta)]$ is almost periodic. The required result follows from application of the Abel-Jacobi-Liouville lemma. This completes the proof of Theorem **6.5**

To complete this section we derive further results from Theorem **6.5** by restricting the kind of exponents allowed in $S(\cup_k \Lambda_{A_k})$. But before doing so we extend a previous result which will be of further use.

Theorem 6.6

Let the exponents $\lambda_j^{(k)}$ of the diagonal matrices $F_k : R \to AP_n$ satisfy the inequality

$$|\lambda_j^{(k)}| \geq M^{(k)} \tag{6.95}$$

and assume that the series

$$F(t,\delta) = \sum_{k=1}^{\infty} F_k(t)\delta^k \tag{6.96}$$

converges uniformly. Then the integral

$$G(t,\delta) = \int_0^t F(s,\delta)ds \tag{6.97}$$

is almost periodic for all $|\delta| < \hat{\rho}$ where

$$\hat{\rho} = \frac{\rho}{\overline{\lim\limits_{k\to\infty}}(dM^{(k)-1})^{k^{-1}}} \tag{6.98}$$

where d is the absolute constant of Lemma **6.2** and ρ is the radius of convergence of (6.96).

Proof: Since the matrix $F(t,\delta)$ is diagonal, we may apply the scalar result given as Lemma **6.3**. In fact we have

$$\sup_t \left| \int_0^t f_{k_{ii}}(s)ds \right| \le \frac{d}{m_i^{(k)}} \| f_{k_{ii}} \| \tag{6.99}$$

where $f_{k_{ii}}(t)$ are the diagonal elements of $F_k(t)$. Therefore

$$\sup_t \left| \int_0^t f_{ii}(s,\delta)ds \right| \le \sum_{k=1}^{\infty} \sup_t \left| \int_0^t f_{k_{ii}}(s)ds\delta^k \right|$$

$$\le \sum_{k=1}^{\infty} \frac{d}{m_i^{(k)}} \| f_{k_{ii}} \| \, |\delta^k| \tag{6.100}$$

Taking the bound

$$M^{(k)} = \min_{1\le i\le n} m_i^{(k)}$$

and applying the Cauchy-Hadamard theorem completes the proof.

Returning to the extension of Theorem **6.5**, we use Theorem **6.6** to prove the following:

Theorem 6.7

Suppose that the almost periodic matrix $A(t,\delta)$ satisfies the conditions of Theorem **6.5**. In addition suppose that all exponents contained in $S(\bigcup_k \Lambda_{A_k})$ are bounded away from zero, so that each $\lambda_j^{(k)} \in \Lambda_{B_k}$ is greater than $M^{(k)}$ in modulus. Then in the circle

$$|\delta| \le \frac{\bar{\rho}}{\overline{\lim\limits_{k\to\infty}}(dM^{(k)-1})^{k^{-1}}} \tag{6.101}$$

$$A(t,\delta) \sim A_o \in \overline{M}_n, \qquad t \in R_+$$

Proof: By hypothesis we observe that in the circle $|\delta| < \bar{\rho} \leq \rho$

$$A(t,\delta) \sim A_0 + \sum_{k=1}^{\infty} B_k(t)\delta^k$$

with

$$\bigcup_k \Lambda_{B_k} \subset S(\bigcup_k \Lambda_{A_k}).$$

Since all exponents of $B_k(t)$ are bounded away from zero for all k also by hypothesis, then according to Theorem **6.6** this implies that

$$\int_0^t \sum_{k=1}^{\infty} B_k(t)\delta^k$$

is almost periodic for

$$|\delta| \leq \frac{\bar{\rho}}{\overline{\lim\limits_{k\to\infty}}(dM^{(k)-1})^{k^{-1}}}$$

It is obvious that the matrices

$$A_0, \quad \int_0^t \sum_{k=1}^{\infty} B_k(t)\delta^k \quad \text{and} \quad [A_0 + \sum_{k=1}^{\infty} B_k(t)\delta^k]$$

all commute for $t \in R_+$ since they are all diagonal. Application of Theorem **4.13** completes the proof.

Note that if the conditions of the theorem are fulfilled and if it is possible to put $\delta = 1$, then the result is similar to Coppel's theorem. The fundamental importance of Theorem **6.7** is seen from the following:

Theorem 6.8: *(Blinov, 1965)*

Let the ordinary differential equation

$$\dot{x}(t,\delta) = A(t,\delta)x(t,\delta) \tag{6.102}$$

possess a matrix of coefficients which satisfies the conditions of Theorem **6.6**. If we suppose that the real parts of the characteristic values of $A_0 \in \overline{M}_n$ are all negative, and that α_j is the characteristic value of least modulus, then the solutions of (6.102) are asymptotically stable for those values of δ for which the inequalities

$$|\alpha_j| > 2 \sum_{k=1}^{\infty} \|A_k\| \, |\delta|^k$$
$$|\delta| < \rho \tag{6.103}$$

hold.

Proof: By hypothesis we observe that in the circle $|\delta| < \bar{\rho} \le \rho$,

$$A(t,\delta) \sim A_o + \sum_{k=1}^{\infty} B_k(t)\delta^k$$

Since the matrices A_o and $B_k(t)$ are diagonal, the simple application of the Gronwall-Bellman inequality to solutions of the linear system kinematically similar to (6.102) shows that they are asymptotically stable for those values of $|\delta| < \bar{\rho} \le \rho$ satisfying

$$|\alpha_j| > \left\| \sum_{k=1}^{\infty} B_k(t)\delta^k \right\|, \quad t \in R_+ \tag{6.104}$$

(see for example Cesari (1971), pp.35-36). However it is easily shown that

$$\left\| \sum_{k=1}^{\infty} B_k(t)\delta^k \right\| \le \sum_{k=1}^{\infty} \|A_k\| \, |\delta|^k +$$
$$\sum_{k=1}^{\infty} \|A_k\| \, |\delta|^k \sum_{k=1}^{\infty} \|P_k\| \, |\delta^k| \tag{6.105}$$

Moreover, $\sum_{k=1}^{\infty} \|P_k\| \, |\delta|^k$ is majorised by $p(\xi) = \sum_{k=1}^{\infty} P_k \xi^k$ and $p(\xi) < 1$ for $|\delta| < \bar{\rho} \le \rho$. Therefore

$$\left\| \sum_{k=1}^{\infty} B_k(t)\delta^k \right\| \le \sum_{k=1}^{\infty} \|A_k\| \, |\delta|^k [1+p(\xi)] \le 2 \sum_{k=1}^{\infty} \|A_k\| \, |\delta|^k \tag{6.106}$$

and the assertion is proved.

Theorem 6.9

Let the ordinary differential equation (6.102) possess a matrix of coefficients $A(t,\delta)$ which satisfies the conditions of Theorem 6.7. If we suppose that the real parts of the characteristic values of $A_o \in \overline{M}_n$ are all negative, then in the circle

$$|\delta| < \frac{\overline{\rho}}{\overline{\lim\limits_{k\to\infty}}(dM^{(k)-1})^{k^{-1}}}$$

the solutions of (6.102) are uniformly asymptotically stable.

Proof: By hypothesis we observe that $A(t,\delta) \sim A_o$ for $|\delta|$ as defined. The assertion follows immediately from the assumptions regarding A_o.

Example 1

Consider the homogeneous system of linear equations

$$\begin{bmatrix} \dot{x}_1(t) \\ \dot{x}_2(t) \end{bmatrix} = \begin{bmatrix} 0 & \dfrac{1}{1-\delta\cos\alpha t} \\ \dfrac{-2}{1-\delta\cos\beta t} & -3 \end{bmatrix} \begin{bmatrix} x_1(t) \\ x_2(t) \end{bmatrix} \tag{6.107}$$

with $\delta < 1$ a real parameter. Equations of this form frequently occur in the study of parametric amplifiers and related devices (Venkatesh, 1977). If α and β are noncommensurable, then the matrix of coefficients of (6.107) is clearly almost periodic. It can be rewritten as

$$A(t,\delta) = \begin{bmatrix} 0 & 1 \\ -2 & -3 \end{bmatrix} + \begin{bmatrix} 0 & \sum\limits_{k=1}^{\infty} \cos^k \alpha t \delta^k \\ -2\sum\limits_{k=1}^{\infty} \cos^k \beta t \delta^k & 0 \end{bmatrix} \tag{6.108}$$

or simply as the series

$$A(t,\delta) = A_o + \sum_{k=1}^{\infty} A_k(t)\delta^k$$

which is convergent for $|\delta| \in [0,1)$ with $A_o \in \overline{M}_2$ and $A_k(t) \in AP_2$ for $k = 1,2,...,$.

By means of a similarity transformation the matrix A_o can be diagonalised; using the same notation we rewrite $A(t,\delta)$ as

$$A(t,\delta) = \begin{bmatrix} -1 & 0 \\ 0 & -2 \end{bmatrix} + \begin{bmatrix} -2\sum_{k=1}^{\infty}(\cos^k\alpha t + \cos^k\beta t)\delta^k & -2\sum_{k=1}^{\infty}(2\cos^k\alpha t + \cos^k\beta t)\delta^k \\ \sum_{k=1}^{\infty}(\cos^k\alpha t + 2\cos^k\beta t)\delta^k & 2\sum_{k=1}^{\infty}(\cos^k\alpha t + \cos^k\beta t)\delta^k \end{bmatrix} \qquad (6.109)$$

Clearly the system of equations (6.107) possess asymptotically stable solutions for $\delta = 0$; for other values of δ stability is difficult to assess. It is also obvious that the conditions of Theorem **6.5** are satisfied so that the matrix $A(t,\delta)$ is kinematically similar to a diagonal matrix $B(t,\delta)$, almost periodic in t for $|\delta| < \bar{\rho} < 1$. We determine $\bar{\rho}$ as follows.

We recall from (6.82) that $\bar{\rho}$ is the smallest positive root of the equation

$$2a(\xi) = 1$$

in which

$$a(\delta) = 2h\sum_{k=1}^{\infty} \|A_k(t)\|\delta^k, \qquad \xi = |\delta|.$$

For this example $\|A_k(t)\| = 10$ for all $k = 1,2,\ldots$ and $h = 1$. Therefore

$$4\sum_{k=1}^{\infty} \|A(t)\| \; |\delta|^k = 1 \qquad (6.110)$$

so

$$40\sum_{k=1}^{\infty} |\delta|^k = \frac{40\xi}{1-\xi} = 1$$

giving $\bar{\rho} = 0.024$ (approximately).

The exact form of the matrix $B(t,\delta)$ is difficult to obtain analytically although the first few terms in the power series

$$B(t,\delta) = B_o + \sum_{k=1}^{\infty} B_k(t)\delta^k \tag{6.111}$$

can be found immediately. In fact

$$B_o = \begin{bmatrix} -1 & 0 \\ 1 & -2 \end{bmatrix} \tag{6.112}$$

and

$$B_1(t) = \begin{bmatrix} -2(\cos\alpha t + \cos\beta t) & 0 \\ 0 & 2(\cos\alpha t + \cos\beta t) \end{bmatrix} \tag{6.113}$$

The evaluation of the remaining terms in the series (6.111) is only feasible by numerical methods. Nevertheless it is possible to determine a little more information about the stability behaviour of system (6.107) by applying Theorem **6.8**. This tells us that the system (6.107) is asymptotically stable for those values of $|\delta| < \bar{\rho}$ for which the inequality

$$|\alpha_j| > 2 \sum_{k=1}^{\infty} ||A_k(t)|| \, |\delta|^k \tag{6.114}$$

is satisfied. Here α_j denotes the negative characteristic value of A_o with least modulus (all characteristic values of A_o must have negative real parts according to the Theorem). Therefore we require that

$$1 > 20 \sum_{k=1}^{\infty} |\delta|^k$$

but this inequality is satisfied for all $|\delta| < \bar{\rho}$, since for $|\delta| = \bar{\rho} = 0.024$, we have

$$20 \sum_{k=1}^{\infty} |\delta|^k = 0.49 < 1.$$

We conclude that the system (6.107) is asymptotically stable for all $|\delta| < \bar{\rho}$.

The same result can be obtained via Theorem **6.9** and Theorem **6.7**. We observe that for α and $\beta \neq 0$, all exponents of $B_k(t)$, $k = 1,2,\ldots$ are bounded away from zero in modulus by some positive constant. Therefore the integral of

$$\sum_{k=1}^{\infty} B_k(t)\delta^k$$

is almost periodic (i.e. bounded) for

$$|\varepsilon| < \frac{\bar{\rho}}{\overline{\lim_{k\to\infty}}(\text{const})^{k^{-1}}} = \bar{\rho} \qquad (6.115)$$

It follows that for $|\delta| < \bar{\rho}$, the matrix $A(t,\delta)$ is kinematically similar to A_o and that system (6.107) is asymptotically stable.

References

Adrianova, L.-Y. (1962). *Vestnik Leningrad Univ.*, **17**, 14-24

Bellman, R. (1953). "Stability theory of differential equations", McGraw-Hill, New York

Blinov, I.N. (1965). *Differentsial'nye Uraveniya*, **1**, 1042-1053

Bohr, H. and Neugebauer, O. (1926). *Nachr.Ges.Wiss.Göttingen, Math.-Physics Klasse*, 8-22

Cesari, L. (1971). "Asymptotic behaviour and stability problems in ordinary differential equations", Springer Verlag, Heidelberg

Coppel, W.A. (1967). *Ann.Mat.Pura Applic.*, **76**, 27-50

Fink, A.M. (1971). *Proc.Am.Math.Soc.*, **27**, 527-533

Fink, A.M. (1974). "Almost periodic differential equations", Lecture Notes in Mathematics, **377**, Springer Verlag, New York.

Gel'man, A.E. (1957). *Dokl.Akad.Nauk.SSSR*, **116**, 535-537

Gel'man, A.E. (1959). *IZV LETI im Ul'yanova (Lenina)*, **39**, 285-293

Gel'man, A.E. (1965). *Differentsial'nye Uraveniya*, **1**, 283-294

Khon, S. (1976). *SIAM Appl.Math.*, **30**, 749-767

Levitan, B.M. (1953). "Almost periodic functions", GIZTTL, Moscow

Lyascenko, N.Ya. (1956). *Dokl.Akad.Nauk.SSSR*, **111**, 295-298

Merkis, V.M. (1968). *Litovsk.Mat.Sb.*, **8**, 101-112

Shtokalo, I.Z. (1946). *Mat.Sb.*, **19**, 236-249

Shtokalo, I.Z. (1960). "Linear differential equations with variable coefficients". *Izd.Akad.Nauk.Ukr.SSSR*, Kiev (in English, Gordon Breach, (1961))

Venkatesh, Y.V. (1977). "Energy methods in time-varying system stability and instability analyses", Lecture notes in physics **No.68**, Springer Verlag, Berlin

Chapter 7

LINEAR SYSTEMS WITH VARIABLE COEFFICIENTS

A surprisingly large number of problems encountered in engineering practice can be formulated in terms of linear differential equations with nonconstant coefficients. In many cases the coefficients depend on time, in others the dependence is spatial. This chapter is intended as a stimulus for applications and further fundamental research based on the contents of the previous six chapters, and to satisfy the need to solve practical problems of an increasingly general nature. We cite several examples of such problems in the introduction and subsequently examine one in some detail. The following survey is certainly not exhaustive, many other examples will be found in the associated literature.

7.1 Introduction and Survey of Applications

The most commonly encountered equations with nonconstant coefficients are those due to Hill and Mathieu. These two equations conveniently describe the behaviour of a great many physical systems and yet they cannot be solved in closed form. Solutions are usually computed numerically using a wide variety of techniques (see for example, Friedmann & Hammond, 1977).

(a) Pendulum with moving support

The equations of motion of a pendulum with moving support are easily derived from the Lagrange equations

$$\frac{d}{dt}\left(\frac{\partial T}{\partial \dot{\theta}}\right) - \frac{\partial T}{\partial \theta} + \frac{\partial V}{\partial \theta} = \Theta$$

$$\frac{d}{dt}\left(\frac{\partial T}{\partial \dot{u}}\right) - \frac{\partial T}{\partial u} + \frac{\partial V}{\partial u} = U$$

where Θ and U are generalised forces,

$$T = \tfrac{1}{2}m[(\ell\dot{\theta}\cos\theta)^2 + (\dot{u}+\ell\dot{\theta}\sin\theta)^2]$$

$$= \tfrac{1}{2}m(\ell^2\dot{\theta}^2 + 2\ell\dot{u}\dot{\theta}\sin\theta + \dot{u}^2)$$

$$V = mg[\ell(1-\cos\theta) + u]$$

and the meaning of other symbols can be deduced from figure 7.1

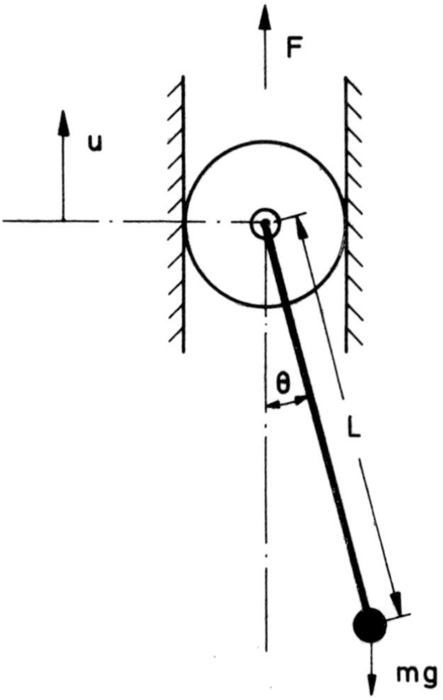

Figure 7.1: A pendulum with moving support

Applying the principle of "virtual work", we conclude that $\Theta = 0$ and $U = F$. Substituting the expressions for T and V into the Lagrange equations and differentiating, we obtain the following scalar nonlinear differential equations:

$$m\ell^2\ddot{\theta} + m\ell\ddot{u}\sin\theta + m\ell g\sin\theta = 0$$

$$m\ell\ddot{\theta}\sin\theta + m\ell^2\dot{\theta}^2\cos\theta + m\ddot{u} + mg = F$$

For θ sufficiently small, these equations reduce to

$$\ddot{\theta} + \ell^{-1}(g + \ddot{u})\theta = 0$$

$$\ddot{u} + g = Fm^{-1}$$

Assume that the displacement of the pivot point is given by $u = a\cos\omega t$. Then the second equation gives

$$F = m(g - a\omega^2\cos\omega t)$$

which represents the force necessary to produce the motion. Furthermore, subject to this assumption the other equation becomes

$$\ddot{\theta} + \ell^{-1}(g - a\omega^2\cos\omega t)\theta = 0$$

For $a = 0$ this is the well known equation of a simple harmonic oscillator. We observe that the equation with $a \neq 0$ admits the equilibrium $\theta = 0$ which is stable for the simple pendulum ($a = 0$). However, it is known that for certain values of $g\ell^{-1}$ and $a\omega^2\ell^{-1}$ the same equilibrium is rendered unstable by moving the pivot support. Setting $g\ell^{-1} = \delta$ and $-a\omega^2\ell^{-1} = 2\xi$, let $\omega = 2$ and $\theta = x$. Then the motion of the pendulum with moving support satisfies the Mathieu equation

$$\ddot{x} + (\delta + 2\xi\cos 2t)x = 0$$

which is studied in detail in McLachlan (1947).

(b) Parametric Amplifiers

Parametric amplification in a simple LC parallel circuit (inductive and capacitive) was one of the first demonstrations of a useful system which can be modelled by a linear differential equation with variable coefficients. Parametric energy transfer is usually achieved by varying periodically the capacitance in the circuit and by choosing the correct frequency the circuit will sustain oscillations. These devices, in a modified form, are used extensively in modern communications systems (Keenan, 1968; MacDonald & Edmundson, 1961; Gardner, 1969; Bell & Wade, 1960).

(c) Clamped-clamped column under periodic axial load

The equation of motion for the clamped-clamped column under periodic axial load shown in Figure 7.2 is (Iwatsubo *et al.*, 1973)

$$m \frac{\partial^2 v}{\partial t^2} + EI \frac{\partial^4 v}{\partial x^4} + (P_1 + P_2 \cos\Omega t) \frac{\partial^2 v}{\partial x^2} = 0$$

and the boundary conditions are

$$v(x,t) = \frac{\partial v(x,t)}{\partial x} = 0 \qquad \text{at} \quad x = 0, L$$

where EI is the bending rigidity. Hamilton's principle can be used to replace the partial differential equation above by a system of coupled Mathieu equations, where the partial properties of

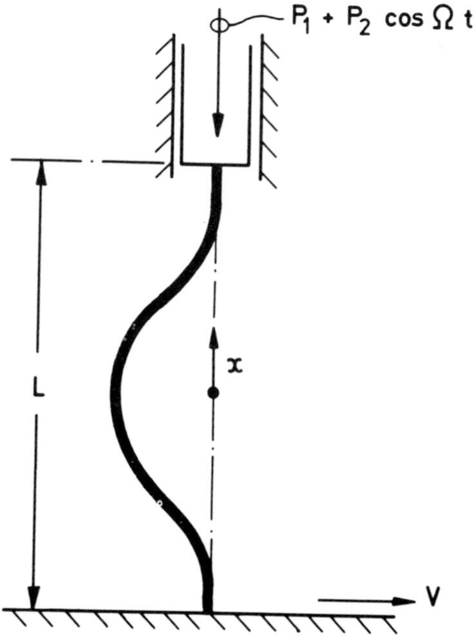

Figure 7.2: Loaded clamped-clamped column

the column are discretised using the finite element methods (Friedmann & Hammond, *op.cit.*).

(d) Electrons in a periodic potential

The quantum mechanical problem of an electron in an electric field that varies periodically with position occurs in a crystal lattice, in which the periodic field is due to the uniformly spaced ions of the lattice. A qualitative view of the problem is obtained if we consider only one dimension. The Schrodinger equation for the wave function $\psi(z)$ of the electron is then

$$\frac{d^2\psi}{dz^2} + 8\pi^2 mh^{-2}[E-V(z)]\psi = 0$$

in which E is the total energy, m is the mass of the electron, $V(z)$ is its potential energy as a result of the electric field and h is Planck's constant. Let α be the spatial period of the electric field (a parameter of the crystal lattice) and write

$$V(z) = V_o - V_1 \cos(2\pi z\alpha^{-1})$$

Setting $\delta = 8\alpha^2 mh^{-2}[E-V_o]$, $\xi = 4\alpha^2 mh^{-2}V_1$, $t = \pi z\alpha^{-1}$ and $\psi = x$ the Schrodinger equation is reduced to the Mathieu equation.

(e) Spacecraft attitude control in the presence of gravity gradient and aerodynamic torques

An inertially referenced cylindrical spacecraft with rectangular solar panels is considered. The vehicle is oriented with its axis of symmetry perpendicular to the orbital plane. Mendel (1968) has shown that the linearised equations of motion of such a spacecraft in circular earth orbit and subject to gravity gradient and aerodynamic torques are

$$\ddot{\psi} = 2\alpha_z^2\psi\cos 2\omega_o t - \alpha_z^2\sin 2\omega_o t$$

$$\ddot{\theta} = \alpha_y^2[\theta(1+\cos 2\omega_o t)-\phi\sin 2\omega_o t]+[\lambda_y^c+\lambda_y^p|\sin(\omega_o t-\psi)|]\sin(\omega_o t-\psi)$$

$$\ddot{\phi} = \alpha_x^2[\phi(1-\cos 2\omega_o t)-\theta\sin 2\omega_o t]+[\lambda_x^c+\lambda_x^p|\sin(\omega_o t-\psi)|]\cos(\omega_o t-\psi)$$

where the inertia tensor I_{ij}, $i,j = x,y,z$ is diagonal, and

$$\alpha_x^2 = 3\omega_o^2(I_{yy}-I_{zz})(2I_{xx})^{-1}$$

$$\alpha_y^2 = 3\omega_o^2(I_{xx}-I_{zz})(2I_{yy})^{-1}$$

$$\alpha_z^2 = 3\omega_o^2(I_{xx}-I_{yy})(2I_{zz})^{-1}$$

$I_{ii}\lambda_i^c$ is the i'th axis component of the aerodynamic torque acting on the cylindrical portion of the spacecraft, $i = x,y,z$

$I_{ii}\lambda_i^p$ is the i'th axis component of the aerodynamic torque acting on the solar panels, $i = x,y,z$

ω_o is the orbital angular velocity

ψ,θ,ϕ is an Euler angle sequence. They define the spacecraft roll, pitch and yaw angles, respectively.

7.2 Beam Stabilisation in an Alternating Gradient Proton Synchrotron

The problem discussed in this section is that of damping transverse oscillations of a proton beam in an alternating gradient (AG) synchrotron. A feedback system is proposed which operates by deflecting the beam at point 2 (refer to Figure **7.3**) by an amount proportional to a detected position error at point 1 (Steining & Wilson, 1974).

The need for some kind of beam "damper" in proton synchrotrons was first recognised when it became possible to inject and accelerate beams whose intensity was above the threshold of *"resistive wall" instability* (Laslett *et al.*, 1965). This effect is essentially an interaction between the proton beam and the metallic walls of the synchrotron vacuum chamber. Although a model for the "resistive wall" instability will not be included in the "damper" model, the response of the "damper" to initial condition type inputs is often used to establish the feedback gain required to suppress the instability once its growth rate is known. Therefore we will concentrate on the problem of designing a "damper" to achieve a prescribed response to given initial conditions

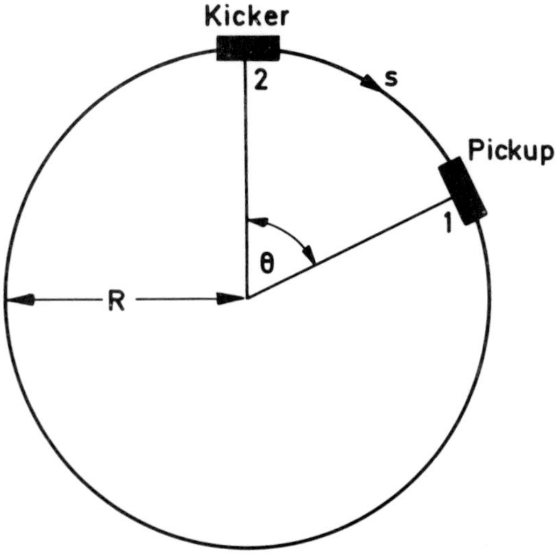

Figure 7.3: Beam deflection in an alternating gradient proton synchrotron

It is not our purpose here to derive in detail the single particle equations of motion in an AG synchrotron, many excellent texts already exist on that subject (Courant & Snyder, 1958; Bruck, 1966). We shall say that the required equations of motion in a circular accelerator are derivable from the relativistic Hamiltonian (Goldstein, 1950)

$$H = e\phi + c[(\mathbf{p}-e\mathbf{A})^2 + m_o{}^2c^2]^{\frac{1}{2}} \tag{7.1}$$

where e is the proton charge, ϕ is a scalar potential of the elec tromagnetic field, **A** is a vector potential of the electromagnetic field, **p** is the canonical momentum of the particle, c is the velo city of light *in vacuo* and m_o is the proton rest mass. In the absence of field imperfections the transverse motion of the proto satisfies the Hill equation

$$\frac{dx(s)}{ds} - A(s)x(s) = 0 \tag{7.2}$$

where

$$A(s) = A(s+2\pi R) = \begin{bmatrix} 0 & 1 \\ -k(s) & 0 \end{bmatrix} \tag{7.3}$$

in which R is the machine radius and

$$k(s) = -\frac{e}{p}\frac{\partial B}{\partial x} \tag{7.4}$$

defines the magnetic focussing strength of the guide field (transverse component). Courant and Snyder show that a formal fundamental matrix for equation (7.2) has the form

$$\Psi(s,s_o) = \begin{bmatrix} \cos\mu(s)+\alpha(s)\sin\mu(s) & \beta(s)\sin\mu(s) \\ -\gamma(s)\sin\mu(s) & \cos\mu(s)-\alpha(s)\sin\mu(s) \end{bmatrix} \tag{7.5}$$

which is usually referred to as the Twiss matrix in accelerator literature. If the differential equation (7.2) possess linearly independent solutions, which will be assumed, then the Wronskian W is equal to the determinant of Ψ (Hale, 1969) and is a constant of the motion; its value $W = 1$ by normalisation. The matrix Ψ can be rewritten as

$$\Phi(s,s_o) = I\cos\mu(s) + J\sin\mu(s) \tag{7.6}$$

where I is the unit matrix and

$$J = \begin{bmatrix} \alpha(s) & \beta(s) \\ -\gamma(s) & -\alpha(s) \end{bmatrix} \tag{7.7}$$

Since $\det\Phi = 1$, we observe that

$$\beta(s)\gamma(s) - \alpha(s)^2 = 1 \tag{7.8}$$

and

$$J^2 = -I \tag{7.9}$$

Assumption 7.1

The parameters α, β and γ of the Twiss matrix (7.5) are real, continuous and bounded on R.

The m'th power of Φ, which is the fundamental matrix for m turns around the machine, is thus

$$\Phi^m = [I\cos\mu(s)+J\sin\mu(s)]^m$$

$$= I\cos m\mu(s)+J\sin m\mu(s) \tag{7.10}$$

It follows from (7.10) that if $\mu(s)$ is real, the elements of Φ^m remain bounded with increasing m, in fact they oscillate. On the other hand, if $\mu(s)$ is not real, $\cos m\mu(s)$ and $\sin m\mu(s)$ become unbounded as $m \to \infty$. Therefore the motion is stable if $\mu(s)$ is rea or

$$|\text{trace } \Phi| \leq 2 \tag{7.11}$$

which is the same thing. Note that by virtue of (7.8) and Assumption 7.1, $\beta(s)$ can never vanish, that is, it is bounded away from zero.

Assumption 7.2

The parameter β of the Twiss matrix (7.5) possesses a real, continuous and bounded first derivative on R.

We now add the feedback structure to the Hill equation as follows:

$$\frac{dx(s)}{ds} = A(s)x(s) + bu(s)\delta(s-2\pi R) \tag{7.12}$$

$$q(s) = cx(R\theta) \tag{7.13}$$

$$u(s) = q(s) \tag{7.14}$$

in which $b = (0 \;\; g)'$, $c = (1 \;\; 0)$, g is the control gain and th azimuth location of the control actuator or *kicker* has been set arbitrarily at $s = 2\pi R$. Equation (7.12) is clearly a forced Hil equation in which the forcing term is impulsive. The measurement equation (7.13), simply states that the beam *position* (transverse coordinate) is measured at the azimuth location $R\theta$ and equation (7.14) defines the feedback law. Combining equations (7.12),

(7.13) and (7.14) yields the following equation for the *closed loop* beam dynamics

$$\frac{dx(s)}{ds} = A(s)x(s) + Ex(R\theta)\delta(s-2\pi R) \quad (7.15)$$

in which $E = bc$. It is possible to reduce equation (7.15) to a more convenient form using the Liapunov transformation

$$P(s) = \begin{bmatrix} \beta(s)^{-\frac{1}{2}} & 0 \\ \frac{-\dot{\beta}(s)\beta(s)^{-\frac{1}{2}}}{2} & \beta(s)^{\frac{1}{2}} \end{bmatrix} \quad (7.16)$$

A diagonal matrix B(s) is obtained which is kinematically similar to A(s). To show this we use the fact (Courant & Snyder, *op cit.*) that β satisfies the equation ($\dot{\beta}(s) \equiv \frac{d\beta}{ds}$ etc.)

$$\frac{\ddot{\beta}(s)\beta(s)}{2} - \frac{\dot{\beta}(s)^2}{4} + k(s)\beta(s)^2 = 1 \quad (7.17)$$

It follows that the closed loop beam dynamics can now be expressed as

$$\begin{aligned} \dot{y}(s) &= P(s)\dot{x}(s) + \dot{P}(s)x(s) \\ &= P(s)A(s)x(s) + P(s)Ex(R\theta)\delta(s-2\pi R) + \dot{P}(s)x(s) \\ &= P(s)A(s)P(s)^{-1}y(s) + P(s)EP(R\theta)^{-1}y(R\theta)\delta(s-2\pi R) + \\ &\quad \dot{P}(s)P(s)^{-1}y(s) \\ &= [P(s)A(s)P(s)^{-1}+\dot{P}(s)P(s)^{-1}]y(s) + \\ &\quad P(s)EP(R\theta)^{-1}y(R\theta)\delta(s-2\pi R) \\ &= B(s)y(s) + P(s)EP(R\theta)^{-1}y(R\theta)\delta(s-2\pi R) \end{aligned} \quad (7.18)$$

where $B(s) \sim A(s)$ is defined by

$$\begin{aligned} B(s) &= P(s)A(s)P(s)^{-1} + \dot{P}(s)P(s)^{-1} \\ &= \beta(s)^{-1}\begin{bmatrix} 0 & 1 \\ -1 & 0 \end{bmatrix} \end{aligned} \quad (7.19)$$

The solution of equation (7.18) can be derived immediately using the variation of constants formula as follows:

$$y(s) = Y(s,0)y(0) + \int_0^s Y(s,\tau)P(\tau)EP(R\theta)^{-1}y(R\theta)\delta(\tau-2\pi R)d\tau \tag{7.20}$$

For $0 \le s < 2\pi R$ we have

$$y(s) = Y(s,0)y(0) \tag{7.21}$$

whereas for $s = 2\pi R$ we have

$$y(2\pi R) = Y(2\pi R,0)y(0) + \int_0^{2\pi R} Y(2\pi R,\tau)P(\tau)EP(R\theta)^{-1}y(R\theta)\delta(\tau-2\pi R)d\tau \tag{7.22}$$

$$\begin{aligned} &= Y(2\pi R,0)y(0) + P(2\pi R)EP(R\theta)^{-1}y(R\theta) \\ &= Y(2\pi R,0)y(0) + P(2\pi R)EP(R\theta)^{-1}Y(R\theta,0)y(0) \\ &= [Y(2\pi R,0)+P(2\pi R)EP(R\theta)^{-1}Y(R\theta,0)]y(0) \\ &= [Y(2\pi R,0)+DY(R\theta,0)]y(0) \end{aligned} \tag{7.23}$$

The fundamental matrix Y is given by

$$Y(s,0) = \begin{bmatrix} \cos\left(\int_0^s \frac{d\sigma}{\beta(\sigma)}\right) & \sin\left(\int_0^s \frac{d\sigma}{\beta(\sigma)}\right) \\ -\sin\left(\int_0^s \frac{d\sigma}{\beta(\sigma)}\right) & \cos\left(\int_0^s \frac{d\sigma}{\beta(\sigma)}\right) \end{bmatrix} \tag{7.24}$$

and $D = P(2\pi R)EP(R\theta)^{-1}$ (7.25)

Using the definition

$$Q = \frac{1}{2\pi}\int_0^{2\pi R} \frac{d\sigma}{\beta(\sigma)} \tag{7.26}$$

it follows that

$$Y(2\pi R,0) = \begin{bmatrix} \cos 2\pi Q & \sin 2\pi Q \\ -\sin 2\pi Q & \cos 2\pi Q \end{bmatrix} \tag{7.27}$$

Clearly the *discrete* closed loop system equation (7.23) defines the beam dynamics over one turn of the machine. The asymptotic behaviour of the beam over many turns is determined by the characteristic values of the matrix

$$S = Y(2\pi R,0) + DY(R\theta,0) \tag{7.28}$$

A well known necessary and sufficient condition for asymptotic stability is that all characteristic values of S lie within the unit circle in the complex plane (Willems, 1970), i.e.

$$|Z_i| < 1, \quad i = 1,2, \quad Z = \text{characteristic values of } S \tag{7.29}$$

However, we note that for S defined by equation (7.28), the same necessary and sufficient condition is expressed by (Kalman & Bertram, p.397, 1960)

$$|\det S| < 1 \tag{7.30}$$

Hence, for asymptotic stability we require that

$$\left| \det \left[\begin{pmatrix} \cos 2\pi Q & \sin 2\pi Q \\ -\sin 2\pi Q & \cos 2\pi Q \end{pmatrix} + \begin{pmatrix} 0 & 0 \\ g\sqrt{\beta_1\beta_2} & 0 \end{pmatrix} \begin{pmatrix} \cos\mu & \sin\mu \\ -\sin\mu & \cos\mu \end{pmatrix} \right] \right| < 1$$

where

$$\mu = \int_0^{R\theta} \frac{d\sigma}{\beta(\sigma)} \quad \text{and} \quad \beta_1 = \beta(R\theta), \quad \beta_2 = \beta(2\pi R). \tag{7.31}$$

By evacuating the determinant it follows that the requirement for asymptotic stability becomes

$$|[1-g\sqrt{\beta_1\beta_2}\,\sin(2\pi Q-\mu)]| < 1 \tag{7.32}$$

Evidently asymptotic stability can be guaranteed by selecting the control gain g, the beta functions at the kicker and pick-up locations and the phase shift μ between the kicker and pick-up to place the characteristic values of the system within the unit circle.

Example

$Q = 26.6$ (value for the C.E.R.N. SPS)

$\mu = 55^{o}$

$\beta_1 = 104\text{m}$ $\qquad g = 10^{-5}\text{rad/cm}$

$\beta_2 = 80\text{m}$

We find that det S = 0.906. Thus the e-folding time (oscillations decrease as $[\det S]^{\frac{m}{2}}$) is approximately 20 revolutions.

References

Bell, C.V. and Wade, G. (1960). *IEEE Trans.Circuit Theory*, **CT-7**, 4-11

Bruck, H. (1966). "Accelerateurs circulaires de particles", Presses Universitaires de France,Paris

Courant, E.D. and Snyder, H.S. (1958). *Ann.Phys.* **3**, 1-48

Friedmann, P. and Hammond, C.E. (1977). *Int.J.Numerical Methods in Engineering* **11**, 1117-1136

Gardner, W.A. (1969). *IEEE Trans.Circuit Theory*, **CT-16**, 295-302

Goldstein, H. (1950). "Classical Mechanics". Addison-Wesley, New York

Hale, J.K. (1969). "Ordinary Differential Equations". Wiley-Interscience, New York

Iwatsubo, T., Saigo, M. and Sugiyama, Y. (1973). *J.Sound and Vibration*, **30**, 65-77

Kalman, R.E. and Bertram, J.E. (1960). *Trans.ASME J.Basic Eng.*

Keenan, R.K. (1968). *Proc.IEEE*, **56**, 1395

Laslett, L.J., Neil, V.K. and Sessler, A.M. (1965). *Rev.Sci.Inst.*, **36**, 437

MacDonald, J.R. and Edmundson, D.E. (1961). *Proc.IRE*, **49**, 453-466

MacLachlan, N.W. (1947). "Theory and Application of Mathieu Functions". Clarendon Press, Oxford

Mendel, J.M. (1968). *IEEE Trans.Automatic Control*, **AC-13**, 362-368

Steining, R. and Wilson, E.J.N., (1974). "Transverse collective instability in the NAL 500 GeV accelerator", *Nuclear Instrum. Methods*, 121, 206-228

Willems, J.L. (1970). "Stability Theory of Dynamical Systems", Nelson, London

Appendix 1

EXISTENCE OF SOLUTIONS TO PERIODIC AND ALMOST PERIODIC DIFFERENTIAL SYSTEMS

A1 Existence Conditions for Periodic Systems

Consider the general homogeneous periodic system

$$\dot{x} = f(t,x), \quad f(t+\omega,x) = f(t,x), \quad \omega > 0 \tag{A1.1}$$

where $f:D \to R^n$ for $D = J \times B$, $J = \{t: |t-t_o| < \alpha,\ t \in R\}$ and $B = \{x: \|x-x_o\| < \beta,\ x \in R^n\}$. Assume that if any solutions to (A1.1) exist then they are unique. By utilising Schauders fixed point theorem (for generalisations appropriate to periodic systems see Browder (1959)), Massera (1950) has shown that if the system (A1.1) is scalar, a solution which exists and remains bounded in the future implies the existence of a *periodic solution* of period ω. Additionally if the system is linear such that

$$f(t,x) = A(t)x + h(t), \tag{A1.2}$$

with $A:J \to M_n$ and $h:J \to R^n$ both continuous and periodic of period ω. Then Massera's result extends to this n-dimension linear periodic system. The proof follows by assuming that $h(t) \neq 0$, then given some initial condition (x_o,t_o) a solution $x_\omega = x(\omega;x_o,t_o)$ through (x_o,t_o) is by (3.57) as

$$x_\omega = X(\omega)X^{-1}(t_o) + X(\omega)\int_0^\omega X^{-1}(s)h(s)ds, \tag{A1.3}$$

where X(t) is the fundamental matrix of the linear homogeneous equation $\dot{x} = A(t)x$. Setting $x_\omega = Gx_o + b \equiv Tx_o$ where G and b are defined by the right hand side of (A1.3) and T is a transformation of x_o; now suppose that the system of linear algebraic equations $(G-I)x+b = 0$ have no solution then $(G-I)$ must be singular and there exists a fixed vector y such that $y'(G-I) = 0$ and $y'b \neq 0$. Since $y'G = y'$ then also $y'G^k = y'$ for $k = 1,2,\ldots,$ and by applying the transformation T repeatedly we have $x_k = T^k x_o = G^k x_o + (G^{k-1}+G^{k-2}+\ldots+I)b$ and hence $y'x_k = y'x_o + ky'b$. But since $y'h \neq 0$ then as $k \to \infty$ $y'x_k \to \infty$ which implies that the solution to (A1.3) is unbounded. But $x_k = x(k\omega;x_o,t_o)$ *is* bounded by definition so by contradiction the linear equation must have a periodic solution.

Consider now the more general case of (A1.1) for n an arbitrary positive integer. By applying Browders (1959) fixed point theorem to the periodic system (A1.1), the following is a generalisation of Cartwright's (1950) result for second order systems:

Theorem A1.1: *Existence of general periodic solution*

If the solutions of (A1.1) are bounded for some bound N, then there exists a periodic solution of period ω such that $||x(t)|| \leq N$ for $t \in R$.

So far an assumption on the uniqueness of solution has been necessary; however by imposing a *stability condition* upon the periodic solution, the uniqueness requirement can be dropped:

Theorem A1.2: *(Yoshizawa, 1975)*

Given that $f(t,x)$ is continuous on D and that the periodic system (A1.1) has a solution $y(t)$ such that $||y(t)|| \leq \beta^* < \beta$ for all $t \in R_+$. Then there exists a periodic solution of (A1.1) of period $r\omega(r \geq 1,2,\ldots)$ if there exists a $N > 0$ such that $||y(t_o)-x_o|| < N$ implies that $||y(t)-x(t;x_o,t_o)|| \to 0$ as $t \to \infty$.

We note that the existence of bounded uniformly asymptotically stable solution to (A1.1) does not necessarily imply the existence of a periodic solution of the same period ω. Continuing the stability approach, Deysach and Sell (1965) have shown that if y(t) is uniformly stable then there exists an *almost* periodic solution to (A1.1); that is we cannot necessarily obtain a periodic solution to a dynamical system with periodic coefficients without imposing additional constraints. Theorem **5.17** has demonstrated that if y(t) is uniformly stable, then y(t) is stable under disturbances from the hull H(f) and so the following theorem follows:

Theorem A1.3: *(Halanay, 1962)*

If the solution y(t) to (A1.1) is uniformly stable then y(t) is *asymptotically* almost periodic and the system (A1.1) has an almost periodic solution which is also uniformly stable.

Finally, Sell (1966) has similarly shown that a periodic solution of period $r\omega$ $(r \geq 1)$ to (A1.1) exists if there is a bounded solution y(t) which is weakly uniformly asymptotically stable (equivalent to uniform asymptotic stability in the case of periodic systems). This result is a special case of Theorem **A1.3** when uniform convergence is used.

A2 Existence Conditions for Almost Periodic Systems

Consider the almost periodic system

$$\dot{x} = f(t,x) \tag{A1.4}$$

where $f{:}R\times D \to R^n$, $(f \in AP(R\times D))$ is almost periodic in t uniformly for $x \in B^* = \{x{:}\ x \in R^n,\ \|x-x_o\| < \beta^*\}$ and for all $t \in R_+$. Let F be a compact subset of B^*, and let y(t) be a solution of (A1.4) such that $\|y(t)\| < \beta^*$, and $y(t) \in F$ for all $t \in R_+$. For some positive sequence $\{a_k\}$ let $\lim_{k\to\infty} f(t+a_k,x) = h(t,x)$ uniformly on $R\times F$, then $h \in H(f)$. Moreover assume that $\lim_{k\to\infty} y(t+a_k) = z(t)$ uniformly on R_+ where z(t) is a solution to

$$\dot{x} = h(t,x), \quad h \in H(f) \tag{A1.5}$$

the dual differential system (A1.4), (A1.5) leads us to the definition of *inherited properties*.

Definition A.1: *Inherited properties of almost periodic systems (Fink, 1972)*

If y(t) has a particular property with respect to system (A1.4), and z(t) has the same property with respect to the dual system (A1.5), then this property is said to be *inherited.*

For almost periodic systems total stability and stability under disturbances are inherited properties; in addition for periodic systems, uniform stability and uniform asymptotic stability are also inherited properties. However it should be noted that for almost periodic systems, uniform stability and uniform asymptotic stability are inherited properties only if the uniqueness of solution is assumed. An example that demonstrates this restriction is given by Kato (1970), it also shows that uniform asymptotic stability does not necessarily imply total stability in almost periodic systems whilst it does for periodic systems.

For periodic systems the boundedness of solution implies the existence of a periodic solution, whilst for almost periodic systems this is not the case. Indeed, several examples of almost periodic systems have been constructed by Opial (1961) and Fink and Frederickson (1971) such that the almost periodic system (A1.4) has no almost periodic solutions yet its solutions are uniformly ultimately bounded. Thus in discussing the existence of almost periodic solutions, stability properties of some kind must be implied. Based upon the assumption of uniqueness of solution to (A1.4), Miller (1965), required that the bounded solution is totally stable for the existence of solution, Seifert (1966) required the Σ-stability of the bounded solution, whilst Sell (1967) required stability under disturbances from the hull (see also section 5.6). All of these results can be achieved without the condition of uniqueness of solution by utilising the property of asymptotically almost periodic functions (see section 2.6). Typical of these results are the following two theorems which are stated without proof:

Theorem A1.4: *Existence of an almost periodic solution (Coppel, 1967)*

If the almost periodic system $\dot{x} = f(t,x)$, $f \in AP(R\times D)$, has a bounded solution on R_+ which is asymptotically almost periodic, then it has an almost periodic solution.

Theorem A1.5: *(Yoshizawa, 1975)*

If the bounded solution y(t) of the almost periodic system (A1.4) is asymptotically almost periodic then for any $h \in H(f)$ there exists a positive sequence $\{a_k\}$ such that $z(t) = \lim_{k\to\infty} y(t+a_k)$ is an almost periodic solution of the system (A1.5) uniformly on R_+.

These theorems tell us that if the almost periodic system (A1.4) has a bounded asymptotically almost periodic solution then the dual system (A1.4-5) also have almost periodic solutions. Also since stability under disturbances from the hull is by Theorem **5.17** a sufficient condition for asymptotic almost periodicity then if the system (A1.4) has a bounded solution which is stable under disturbances from H(f), then it also has an almost periodic solution y(t) which is also stable under disturbances from H(f). A corollary to this result is that if the above solution y(t) is totally stable then y(t) is asymptotically almost periodic and the system (A1.4) has an almost periodic solution which is totally stable.

In the previous section we have seen that if a periodic system has a bounded and stable solution then there exists an almost periodic solution. Examples of periodic systems can be generated (Yoshizawa, 1975) whereby the system has a quasi-periodic solution and thus the module of the almost periodic solution is *not* contained within the module of the system. This demonstrates that uniform stability and stability under disturbances from the hull do not give module containment for almost periodic systems. In the following we develop the conditions, based upon uniform stability of solutions, for the existence of almost periodic solutions to the almost periodic linear inhomogeneous system:

$$\dot{x} = A(t)x + f(t), \tag{A1.6}$$

where $A:R \to M_n$ and $f:R \to R^n$ are almost periodic uniformly in t on R. Corresponding to (A1.6) is the homogeneous linear system

$$\dot{x} = A(t)x, \tag{A1.7}$$

and the equation on the hull

$$\dot{x} = G(t)x, \qquad G \in H(A). \tag{A1.8}$$

Clearly if the almost periodic system (A1.6) has a bounded solution on R_+ which is uniformly stable then the null solution of (A1.7) and (A1.8) are also uniformly stable. Under these conditions the following theorem due to Favard (1933) shows that (A1.6) has an almost periodic solution:

Theorem A1.6

If every nontrivial solution x(t) of (A1.8) on the homogeneous hull of the linear almost periodic system (A1.6) is bounded on R and satisfies $\inf_{t \in R} \|x(t)\| > 0$, then if (A1.6) has a bounded solution on $t \in R_+$, there exists an almost periodic solution y(t) of (A1.6) such that $\text{mod}(y) \subset \text{mod}(A,f)$, where mod(y) is the module of y and mod(A,f) is the set module of A on f.

Proof: (Outline) It is straightforward to show that a bounded solution $y_0(t)$ to (A1.6) exists and each equation in the hull of (A1.6) has a unique solution with minimum norm. It is then a simple matter to show that $y(t), y_0(t) \in AP(R \times R^n)$, and finally by module containment (see definition **2.4**) that $\text{mod}(y) \subset \text{mod}(A,f)$.

By similar reasoning Bochner (1962) derived as a corollary to Farvard's theorem (A1.6):

Theorem A1.7

If every equation in the homogeneous hull of the linear almost periodic system (A1.6) is almost periodic, then every bounded solution of (A1.6) is almost periodic.

Bochner's result also holds for a variety of important special cases; (i) the Bohr-Neugebauer theorem (1926) for linear constant coefficient almost periodic systems ($A \in \overline{M}_n$, $f \in AP(R^n)$), (ii) when $f(t) \equiv 0$ and $A(t) \in AP(M_n)$, and (iii) when $A(t)$ is purely periodic, that is $A(t+\omega) = A(t)$ for some $\omega > 0$.

References

Bochner, S. (1962). *Proc.Nat.Acad.Sci.(USA)* **48**, 2039-2043

Bohr, H. and Neugebauer, O. (1926). "Uber lineare differentialgleichungen mit konstanten koeffizienten und fastperiodischer rechter seite", *Nachr.Ges.Wiss.Göttingen, Math.-Phys.Klass.*, 8-22

Browder, F.E. (1959). *Duke Math.J.* **26**, 291-303

Cartwright, M.L. (1950). "Forced oscillations in nonlinear systems", contri. to "The theory of nonlinear oscillations", Ed. S. Lefschetz **1**, Princeton University Press

Coppel, W.A. (1967). *Ann.Mat.Pura Applic.* **76**, 27-50

Deysach, L.G. and Sell, G.R. (1965). *Michigan Math.J.* **12**, 87-95

Favard, J. (1933). "Leçons sur les Fonctions Presque Périodiques", Gauthier Villars, Paris

Fink, A.M. (1972). *SIAM Review* **14**, 572-581

Fink, A.M. and Frederickson, P.O. (1971). *J.Diff.Eqns.* **9**, 280-284

Halanay, A. (1962). *Uspeh Mat.Nauk.* **17**, 231-233

Kato, J. (1970). *Tohoku Math.J.* **22**, 254-269

Massera, J.L. (1950). *Duke Math.J.* **17**, 457-475

Miller, R.K. (1965). *J.Diff.Eqns.* **1**, 293-305

Opial, Z. (1961). *Bull.Acad.Polon.Sci.Ser.Sci.Math.Astron.Phys.* **9**, 673-676

Seifert, G. (1966). *J.Diff.Eqns.* **2**, 305-319

Sell, G.R. (1966). *J.Diff.Eqns.* **2**, 143-157

Sell, G.R. (1967). *Trans.Math.Soc.* **127**, 241-283

Yoshizawa, T. (1975). "Stability theory and the existence of periodic solutions and almost periodic solutions". Appl.Maths.Sci. No.14, Springer Verlag, New York

Appendix 2

DICHOTOMIES AND KINEMATIC SIMILARITY

The concepts of exponential dichotomies and kinematic similarity were respectively introduced in Chapters 5 and 4. We now demonstrate that kinematically similar systems satisfy common dichotomies. Unfortunately the problem of establishing that the existence of a dichotomy implies kinematic similarity is not completely resolved, although some results are available.

Definition A2.1: *Kinematically Similar Systems*

The linear autonomous differential systems $\dot{x} = A(t)x$, $\dot{y} = B(t)y$, with $A(t), B(t) \in M_n$ are said to be *kinematically similar systems* if $A(t) \sim B(t)$.

Our first formal observation is that two kinematically similar systems satisfy common exponential dichotomies.

Theorem A2.1

If a homogeneous linear equation has an exponential dichotomy then any equation kinematically similar to it likewise has an exponential dichotomy with the same projection P_o and the same constants α, β.

Proof: If $\dot{y}(t) = B(t)y(t)$ is obtained from $\dot{x}(t) = A(t)x(t)$ by the change of variables $x(t) = P(t)y(t)$, where $P(t)$ is a Liapunov transformation, then $Y(t) = P(t)^{-1}X(t)$ and

$$|Y(t)P_oY(s)^{-1}| \le \text{const}|X(t)P_oX(s)^{-1}|$$

$$|Y(t)(I-P_o)Y(s)^{-1}| \quad \le \quad \text{const}\,|X(t)(I-P_o)X(s)^{-1}|$$

The assertion then follows trivially.

The interesting problem is to show that a system satisfying an exponential dichotomy is kinematically similar to some other system. One result for this is due to Coppel (1967) and is concerned with conditionally stable systems (the same result for non-conditionally stable systems is trivial). First of all we require the following:

Lemma A2.1

Let P_o be a projection and let $X(t)$ be a continuous nonsingular matrix such that $X(t)P_oX(t)^{-1}$ is bounded for all t. Then there exists a continuous nonsingular matrix $P(t)$ such that

$$P(t)P_oP(t)^{-1} \quad = \quad X(t)P_oX(t)^{-1} \tag{A2.1}$$

which is bounded, together with its inverse for all t.

Proof: Suppose, without loss of generality, that P_o is an orthogonal projection, that is $P_o = P_o^*$. Since any positive Hermitian matrix has a unique positive square root, there exists, for each t, a unique $R(t) = R(t)^* > 0$ such that

$$R(t)^2 \quad = \quad P_oX(t)^*X(t)P_o \; + \; (I-P_o)X(t)^*X(t)(I-P_o) \tag{A2.2}$$

Moreover, since $R(t)^2$ commutes with P_o, so does $R(t)$. Let $|\text{matrix}|^2 = \text{tr}(\text{matrix}^* \text{ matrix})$. It follows at once from the definition of $R(t)$ that

$$n \quad = \quad |X(t)P_oR(t)^{-1}|^2 \; + \; |X(t)(I-P_o)R(t)^{-1}|^2$$

and hence

$$|X(t)R(t)^{-1}| \quad \le \quad |X(t)P_oR(t)^{-1}| + |X(t)(I-P_o)R(t)^{-1}| \quad \le \quad (2n)^{\frac{1}{2}} \tag{A2.3}$$

On the other hand, if $|X(t)P_oX(t)^{-1}| \le \eta$, then $|X(t)(I-P_o)X(t)^{-1}| \le \xi$, where $\xi = \eta + n^{\frac{1}{2}}$. Thus

$$|R(t)X(t)^{-1}|^2 \quad = \quad |X(t)P_oX(t)^{-1}|^2 + |X(t)(I-P_o)X(t)^{-1}|^2$$

gives

$$|R(t)X(t)^{-1}| \le (\eta^2 + \xi^2)^{\frac{1}{2}} \tag{A2.4}$$

Finally R(t) is continuous. The lemma now follows at once if we take $P(t) = X(t)R(t)^{-1}$.

Notice that if $P_o = 0$ or 1, then $R(t) = X(t)$ and $P(t) = I$. The result of importance is:

Theorem A2.2

Given the homogeneous system $\dot{x}(t) = A(t)x(t)$ with $A(t) \in M_n$, suppose that there exists a projection P_o such that $X(t)P_oX(t)^{-1} \in M_n$. Then the given system is kinematically similar to the system $\dot{y}(t) = B(t)y(t)$ with $B(t) \in M_n$ such that $P_oB(t) = B(t)P_o$ for all t.

Proof: Subject to the conditions of the theorem, the matrix functions R(t) and P(t) of the previous lemma are continuously differentiable. The change of variables $x(t) = P(t)y(t)$ defines the kinematic similarity where

$$B(t) = -P(t)^{-1}[\dot{P}(t)-A(t)P(t)].$$

Since the new system has R(t)as a f.m., $B(t) = \dot{R}(t)R(t)^{-1}$ commutes with P_o.

Because A(t) is bounded, there exists a positive constant θ such that for every t

$$-\theta I \le A(t) + A^*(t) \le \theta I \tag{A2.5}$$

If $P_o = P_o^*$, then from the definition of R(t) we have

$$R(t)\dot{R}(t) + \dot{R}(t)R(t) = P_oX(t)^*(A(t) + A(t)^*)X(t)P_o$$
$$+ (I-P_o)X(t)^*(A(t) + A(t)^*)X(t)(I-P_o) \tag{A2.6}$$

It follows that

$$-\theta R(t)^2 \le R(t)\dot{R}(t) + \dot{R}(t)R(t) \le \theta R(t)^2 \tag{A2.7}$$

and hence

$$-\theta I \le \dot{R}(t)R(t)^{-1} + R(t)^{-1}\dot{R}(t) \le \theta I \tag{A2.8}$$

Therefore

$$\{\dot{R}(t)R(t)^{-1} + R(t)^{-1}\dot{R}(t)\}^2 \leq \theta^2 I \tag{A2.9}$$

and

$$|\dot{R}(t)R(t)^{-1} + R(t)^{-1}\dot{R}(t)| \leq \theta_n^{\frac{1}{2}} \tag{A2.10}$$

We are going to deduce that $B(t) = \dot{R}(t)R(t)^{-1}$ is bounded.

Suppose that G and H are Hermitian matrices and $G \geq 0$. Since

$$(GH + HG)^2 = (GH)^2 + GH^2G + HG^2H + (HG)^2$$

and the trace of a product is unaltered by cyclic permutation of the factors, then

$$\mathrm{tr}(GH + HG)^2 = 2\mathrm{tr}(GH^2G) + 2\mathrm{tr}(GH)^2.$$

But $G = S^2$ for some Hermitian matrix S and hence

$$\mathrm{tr}(GH)^2 = \mathrm{tr}(SHS)^2 = |SHS|^2 \geq 0 \tag{A2.11}$$

Therefore

$$|HG|^2 = \mathrm{tr}(GH^2G) \leq \tfrac{1}{2}\mathrm{tr}(GH + HG)^2 = \tfrac{1}{2}|GH + HG|^2 \tag{A2.12}$$

Thus (A2.10) implies that

$$|B(t)| \leq \theta[\tfrac{1}{2}n]^{\frac{1}{2}}$$

and the assertion is proven.

We observe that the form of the matrix B(t) in the above theorem is

$$B(t) = \begin{pmatrix} B_1(t) & 0 \\ 0 & B_2(t) \end{pmatrix} \tag{A2.13}$$

where $B_1(t)$ and $B_2(t)$ are matrices of lower order than B(t). These correspond to two sets of solutions of the equation $\dot{y}(t) = B(t)y(t)$ with different exponential growth behaviour. For $P_o = 0$ or I we have the trivial case $A(t) = B(t)$ with $B(t) = B_1(t)$ or $B_2(t)$.

Theorems **A2.1** and **A2.2** can be combined as follows:

Theorem A2.3

If a homogeneous system possesses an exponential dichotomy with projection P_o then it is kinematically similar to a system which also possesses an exponential dichotomy and whose coefficient matrix commutes with P_o.

We note that theorems **A2.1**, **A2.2** and **A2.3** are equally applicable to systems possessing an ordinary dichotomy. Other results on the relationship between dichotomies and kinematic similarity are contained in the papers of Sacker and Sell (1974), Sell (1974) and Coppel (1968). In particular Coppel *(op.cit.)* considers systems with almost periodic coefficients; further results on kinematic similarity in this context can be found in Fink (1974), and in Chapter 6.

We have already noted in section 5.3 the relationship between stability and dichotomies; this suggests that the existence of a Liapunov function implies an exponential dichotomy. To see this equivalence consider the linear differential equation

$$\dot{x} = A(t)x, \qquad A(t) \in M_n \tag{A2.14}$$

If X(t) is the fundamental solution of (A2.14) with $X(o) = I$, define the Hermitian matrix function

$$\begin{aligned} Q(t) = {} & 2\int_t^{\infty} (X(t)P_oX(s)^{-1})^*(X(t)P_oX(s)^{-1})ds \\ & - 2((I-P_o)X(t)^{-1})^*(I-P_o)X(t)^{-1} \\ & - 2\int_o^t (X(t)(I-P_o)X(s)^{-1})^*(X(t)(I-P_o)X(s))ds \end{aligned} \tag{A2.15}$$

Assume that (A2.14) has an exponential dichotomy (see (5.22)) then

$$||Q(t)|| \le 2\ell k(\alpha+\beta+1)(\alpha+\beta)^{-1} \tag{A2.16}$$

which provides an upper bound to Q(t).

Now since $\dot{X} = A(t)X$ then from (A2.15),

$$\dot{Q} + QA(t) + A(t)^*Q = -2R^*R - 2(I-R)^*(I-R) \tag{A2.17}$$

where $R \equiv X(t)P_o X(t)^{-1}$. But in view of the properties of the projection operator P_o (see section 5.3) the right hand side of (A2.17) bounds the identity matrix, therefore

$$\dot{Q} + QA(t) + A(t)^*Q \leq -I \qquad \text{for} \quad t \geq 0. \tag{A2.18}$$

Clearly if $||A(t)|| < \infty$ then $||\dot{Q}|| < \infty$ since $||Q|| < \infty$.

Finally we note that the quadratic function (a Liapunov function)

$$V = x^*Qx$$

has a negative definite derivative

$$\dot{V} = x^*[Q + QA(t) + A(t)^*Q]x \leq -||x||^2 \tag{A2.19}$$

We conclude that an exponential dichotomy for (A2.14) is sufficient for the existence of a quadratic Liapunov function with a negative definite derivative. The converse is not true unless the solution to (A2.14) is bounded. The connection between Liapunov functions and exponential dichotomies was first noted by Maizel (1954) and later by Massera and Schaffer (1966).

References

Coppel, W.A. (1967). *J.Differential Eqns.*, **3**, 500-521

Coppel, W.A. (1968). *J.Differential Eqns.*, **4**, 386-398

Fink, A.M. (1974). "Almost periodic differential equations", Lecture Notes in Mathematics No.377, Springer Verlag, New York

Maizel, A.D. (1954). *Ural.Politehn.Inst.Trudy*, **51**, 20-50

Massera, J.L. and Schaffer, J.J. (1966). "Linear differential equations and function spaces", Academic Press, New York.

Sacker, R.J. and Sell, G.R. (1974). *J.Differential Eqns.*, **15**, 429-458

Sell, G.R. (1974). "The Floquet problem for almost periodic linear differential equations", Lecture Notes in Mathematics, No.415, Springer Verlag, New York

Appendix 3

BIBLIOGRAPHY

The papers here are collected in three sections; on Differential Equations with Almost Periodic Coefficients, on Reducibility and Kinematic Similarity, and on Stability.

(i) References on Differential Equations with Almost Periodic Coefficients

Abel, J. (1970). On the almost periodic Mathieu equation, *Quart. J.Appl.Math.* **28**, 205-217

Amerio, L. (1967). Almost periodic solutions of the equation of the Schrodinger type. *Atti Acad.Nay Lincei Rend.Cl.Sci.Fiz.Mat. Natur.* **8**, 147-153

Amerio, L. and Prouse, G. (1971). Almost periodic functions and functional equations. Van Nostrand, New York

Artjusenko, L.M. (1968). The application of Fourier series for finding almost periodic solutions of equations with mean values. *Izv.Vyss.Ucebn.Zaved Matematika* **5**, 21-27

Barbalat, I. (1961). Solutions presque-periodiques des equations differentielles non-linearies. *Com.Acad.R.S.Romainia* **11**, 155-159

Berezanskii, Y.M. (1953). On generalised almost periodic functions and sequences, related with the difference-differential equations. *Mat.Sb.* **32**, 157-194

Birjuk, G.I. (1954). On a theorem concerning the existence of almost periodic solutions for certain nonlinear differential systems with a small parameter. *Doklady.Akad.Nauk.SSSR* **96**,5-7

Blinov, I.N. (1965). Analytic representation of the solution of a system of linear differential equations with almost periodic coefficients which depend upon a parameter. *Diff.Uravnenija* **1**, 1042-1053

Blinov, I.N. (1965). An analytical solution of a linear system of differential equations with periodic coefficients depending upon

a parameter. *Diff.Equations* **1**, 679-691, 812-821
Blinov, I.N. (1971). The regularity of a class of linear systems with almost periodic coefficients. *Diff.Equations* **3**, 764-768
Bochner, S. (1933). Homogeneous systems of differential equations with almost periodic coefficients. *J.London Math.Soc.* **8**, 283-288
Bogdanowicz, W.M. (1963). On the existence of almost periodic solutions for systems of ordinary differential equations in Banach spaces. *Arch.Rational Mech.Anal.* **13**, 364-370
Bohr, H. (1952). Collected Mathematical Works, Vols.I, II, III, *Dansk.Mat.Forening*, Kobenhavn
Boruhov, L.E. (1947). A linear integral equation with almost periodic kernel and free term. *Doklady.Akad.Nauk.SSSR.* **57**, 647-649
Boruhov, L.E. (1954). On almost periodicity of solutions of some linear differential systems with almost periodic coefficients. *Nauk.Ez.Saratovsk.Univ.* 659-660
Burd, V.S. (1965). Dependence on a parameter of almost periodic solutions of differential equations with a deviating argument. *Akad.Nauk.Azerbaidzen.SSR.Dokl.* **21**, 3-7
Burton, T.A. (1966). Linear differential equations with periodic coefficients. *Proc.Amer.Math.Soc.* **17**, 327-329
Bylov, B.F. (1965). The structure of the solutions of a system of linear differential equations with almost periodic coefficients. *Mat.Sb.* **66**, 215-229
Bystrenin, V.V. (1941). On almost periodic solutions of certain ordinary differential equations. *Doklady.Akad.Nauk.SSSR* **33**, 387-389
Cameron, R.H. (1938). Linear differential equations with almost periodic coefficients. *Acta Math.* **6**, 21-56
Cartwright, M.L. (1971). Almost periodic solutions of differential equations and flows. *Global differential Dynamics, Proc.Conf. Case Western Univ., U.S.A.*. Springer Verlag Lecture Notes, Math. **235**, 35-43
Chang, K.W. (1968). Almost periodic solutions of singularly perturbed systems of differential equations. *J.Diff.Equations* **4**, 300-307
Coppel, W.A. (1967). Almost periodic properties of ordinary differential equations. *Ann.di Mat.Pura ed Appl.* **76**, 27-50
Conti, R. and Sansone, G. (1964). Nonlinear differential equations. Pergamon, New York
Corduneanu, C. (1968). Almost periodic functions. Interscience Publishers, New York.
Doss, R. (1965). On the almost periodic solutions of a class of integro-differential-difference equations. *Ann.Math.* **81**, 117-123
Emzarov, K. and Tulegenov, M. (1966). An existence theorem for almost periodic solutions of a differential equation with a small parameter in a Banach space. *Vestnik Akad.Nauk Kazah. SSR* **22**, 42-44
Erugin, N.P. (1966). Linear systems of ordinary differential

equations with periodic and quasi-periodic coefficients. Academic Press, New York.

Ezeilo, J.O.C. (1966). On the existence of almost periodic solutions of some dissipative second order differential equations. *Ann.Mat.Pura Appl.* **74**, 399

Favard, J. (1933). Leçons sur les fonction Presque-périodiques. Gauthier-Villars, Paris

Favard, J. (1963). Sur certains systèmes différentiels scalaires linéaires et homogènes à coefficients presque-periodiques. *Ann.Mat.Pura Appl.* **4**, 61

Fink, A.M. (1972). Almost periodic functions invented for specific purposes. *SIAM Review* **14**, 572-581

Fink, A.M. and Frederickson, P. (1971). Ultimate boundedness does not imply almost periodicity. *J.Diff.Equations* **9**, 280-284

Fink, A.M. and Seifert, G. (1971). Nonresonance conditions for the existence of almost periodic solutions of almost periodic systems. *SIAM J.Appl.Math.* **21**, 362-366

Frechet, M. (1941). Les fonctions asymptotiquement presque periodiques, *Rev.Scientifique* **79**, 341-354

Goldberg, R.R. (1957). Convolutions transforms of almost periodic functions. *Riv.Mat.Univ.Parma.* **8**, 307-312

Golomb, M. (1958). Expansions and boundedness theorems for solutions of linear differential systems with periodic or almost periodic coefficients. *Arch.Rat.Mech.Analysis* **2**, 284-308

Gunzler, H. and Zaidman, S. (1969). Abstract almost periodic differential equations. Abstract Spaces and Approximation. Birkhauser, Basel, 387-392

Gurjanov, A. (1970). On sufficient conditions for the regularity of second order systems of ordinary differential equations with uniform asympotically almost periodic coefficients. *Vestnick Leningrad Univ.* **25**, 23-27

Halanay, A. (1960). Almost periodic solutions of systems of differential equations with a lagging argument and a small parameter. *Rev.Math.pures et Appl.* **5**, 75-79

Halanay, A. (1963). Almost periodic solutions of systems with a small parameter in a certain critical case. *Rev.Math.pures et Appl.* **8**, 397-403

Hale, J.K. (1964). Periodic and almost periodic solutions of functional differential equations. *Arch.Rat.Mech.Anal.* **15**, 289-304

Hale, J.K. (1969). Ordinary differential equations. Wiley-Interscience, New York.

Harasahal, V.H. (1960). Almost periodic solutions of nonlinear systems of differential equations. *Prikl.Mat.Meh.* **24**, 565-567

Hermes, H. (1973). A survey of recent results in differential equations. *SIAM Review* **15**, 453-468

Ivanov, V.N. (1965). On linear differential operators in the space of almost periodic functions. *Trudy Saratovsk.Inst.Meh.Selsk.* **38**, 141-149

Jakubovic, V.A. (1966). Periodic and almost periodic limit regimes of automatic control systems with some discontinuous nonlinearities. *Doklady Akad.Nauk.SSSR* **171**, 533-537

Kapisev, K.K. (1966). Quasiperiodic solutions of nonlinear systems of differential equations containing a small parameter. *Vestnik Akad.Nauk Kazah.SSR* **22**, 42-47

Kempner, G.A. (1968). Almost periodic functional differential equations. *SIAM J.Appl.Math.* **16**, 155-161

Kovanko, A.S. (1965). Almost periodic solutions of certain differential equations with almost periodic right sides. *Visnik Lwow.Univ.Ser.Math.fasc.* 2, 3-8

Krasnoselskii, M.A. and Perov, A.I. (1958). A principle concerning the existence of bounded periodic and almost periodic solutions for systems of ordinary differential equations. *Doklady Akad. Nauk.SSSR* **123**, 235-238

Langenhop, C.E. and Seifert, G. (1959). Almost periodic solutions of second order nonlinear differential equations with almost periodic forcing. *Proc.Am.Math.Soc.* **10**, 425-432

Levitan, B.M. (1937). On linear differential equations with almost periodic coefficients. *Doklady Akad.Nauk.SSSR* **17**, 285-286

Lillo, J.C. (1959). On almost periodic solutions of differential equations. *Ann.Math.* 69, 467-485

Lisevic, L.N. (1960). Extension of Favard's theorems to the case of a linear system of differential equations with analytic almost periodic coefficients. *Dovopidi Akad.Nauk.Ukrain.SSR*, 148-149

Ljubarskii, M. (1972). The extension of Favard's theory to the case of a system of linear differential equations whose coefficients are unbounded and almost periodic in the sense of Levitan. *Soviet Math.* **13**, 1316-1319

Lyascenko, N.Ya. (1956). An analogue of the theorem of Floquet for a special case of linear homogeneous systems of differential equations with quasi-periodic coefficients. *Doklady Akad.Nauk. SSSR* **111**, 295-298

Malkin, I.G. (1954). On almost periodic oscillations of nonlinear non-autonomous systems. *Prikl.Mat.Mech.* **18**, 681-704

Marcus, L. and Moore, R.A. (1956). Oscillations and disconjugacy for linear differential equations with almost periodic coefficients. *Acta.Math.* **96**, 99-123

Massera, J.L. and Schaffer, J.J. (1958). Linear differential equations and functional analysis. *I.Ann.Math.* **67**, 517-572

Miller, R.K. (1965). Almost periodic differential equations as dynamic systems with applications to the existence of almost periodic solutions. *J.Differential Equations* **1**, 337-345

Millionscikov, V.M. (1965). Recurrent and almost periodic limit trajectories of nonautonomous systems of differential equations. *Doklady Akad.Nauk.SSSR* **161**, 43-44

Mitropolskii, Y.A. and Samoilenko, A.M. (1965). On constructing solutions of linear differential equations with quasi-periodic coefficients by the method of improved convergence. *Ukrain.Mat. Z.* **17**, 42-59

O'Brien, G.C. (1972). Almost periodic and quasi-periodic solutions of differential equations. *Bull.Aust.Math.Soc.* **7**, 453-454

Oleinik, S.G. (1969). The investigation of linear systems of differential equations with almost periodic coefficients. *Math.Phys. Nauk.Dumka, Kiev.*, **6**, 139-149

Opial, Z. (1961). Sur une equation differentieble presque-periodique sans solution presque-periodique. *Bull.Acad.Polon.Sci. Ser.Sci.Math.Astr.Phys.* **9**, 673-676

Phillips, R.S. (1940). On linear transformations. *Trans.Am.Math. Soc.* **48**, 516-541

Ragimov, M.B. and Zadoroznii, V.G. (1970). Almost periodic solutions of multidimensional differential equations. *Akad.Nauk. Azerbaidzen.SSR.Dokl.* **26**, 8-11

Rjabov, J.A. (1963). On a method of finding a bound for the region of existence of periodic and almost periodic solutions of quasilinear differential equations with a small parameter. *Izv.Vyss. Ucebn.Zaved.Matematika* **33**, 101-107

Sanchez, D.A. (1969). A note on periodic solutions of Riccati-type equations. *SIAM J.Appl.Math.* **17**, 957

Sansone, G. and Conti, R. (1964). Nonlinear differential equations, Pergamon Press, London

Seifert, G. (1966). Almost periodic solutions for almost periodic systems of ordinary differential equations. *J.Diff.Eqns.* **2**, 305-319

Seifert, G. (1972). Almost periodic solutions for limit periodic systems. *SIAM J.Appl.Math.* **22**, 38-44

Shtokalo, I.Z. (1960). Linear differential equations with variable coefficients. *Izd.Akad.Nauk.Ukr.SSSR, Kiev.* (also Gordon-Breach, 1961)

Svinbekov, K.D. (1964). On the analytic form of solutions of linear systems of differential equations with quasi-periodic coefficients. *Izv.Akad.Nauk.Kazah.SSR Ser.Fiz.Mat.Nauk.* **2**, 69-71

Talpalaru, P. (1969). Solutions periodiques et presque-periodiques des systems differentiels. *An.sti.Univ.Al.I.Cunza Iasi n Ser Sect.Ia* **15**, 375-385

Turcu, A. (1965). Almost periodic solutions of the equations of Duffing in the case of resonance. *Studia Univ.Babes Bolyaik. Ser.Math.-Phys.* **10**, 83-94

Umbetzanov, D.U. (1970). The almost periodic solution of certain classes of partial differential equations with small parameters. *Differencial nye Uravenija* **6**, 913-916

Urabe, M. (1972). Existence theorems of quasi-periodic solutions to nonlinear differential systems. *Funk.Ekv.* **15**, 75-100

Vaghi, C. (1968). Soluzioni limitate, o quasi-periodiche, di un' equazione di tipo parabolico nonlineare. *Boll.U.M.I.* **4-5**, 559-580

Valeev, K. (1969). Linear differential equations with quasi-periodic coefficients and constant retardation of the argument. *Visnik Kiev Univ.Ser.Mat.Mech.*, 16-24

Vereennilov, V. (1969). On the construction of solutions of quasi-linear nonautonomous systems in resonance cases. *Appl.Math.Mech.* **33**, 1090-1099

Vzovskii, D.A. (1972). The almost periodic solutions of certain nonlinear systems of differential equations with deviating argument. *Differencial'nye Uravnerija* **8**, 415-423

Wexler, D. (1966). Solutions periodiques et presque-périodiques des systèmes d'equations différentielles linéaires en distributions. *J.Diff.Eqns.* **2**, 12-32

Yoshizawa, T. (1969). Asymptotically almost periodic solutions of an almost periodic system. *Funk.Ekv.* **12**, 23-40

Zaicer, A.I. (1967). Analytic form of solutions of linear systems of differential equations with quasi-periodic coefficients. *Differencial'nye Uravnenija* **3**, 219-225

Zaicer, A.I. and Kapish, K. (1967). Quasiperiodic solutions of non-linear systems of differential equations containing a small parameter. *Diff.Eqns.* **7**, 1100-1112

Zaidman, S. (1969). Spectrum of almost periodic solutions for some abstract differential equations. *J.Math.Anal.Appl.* **28**, 336-338

Zikov, V.V. (1970). Almost periodic solutions of linear and non-linear equations in a Banach space. *Soviet Math.* **11**, 1457-1461

Zubov, V.I. (1960). Periodic and almost periodic forced oscillations arising from the action of an external force. *Izv.Vyss. Ucebn.Zaved.Mat.* **19**, 93-102

(ii) References on Reducibility and Kinematic Similarity

Adrianova, L.J. (1962). The reducibility of systems of n-linear differential equations with quasi-periodic coefficients. *Vestnik Leningrad Univ.* **17**, 14-24

Alimzanova, R.M. and Zolotarex, J.G. (1966). The transformation of analytic systems of differential equations with quasi-periodic coefficients. *Kazah.Gos.Ped.Inst.Ucen.Zap.* **23**, 45-54

Blinov, I.N. (1965). An analytic solution of a linear system of differential equations with periodic coefficients depending upon a parameter. *Diff.Uravneniye* **1**, 880-891

Bylov, B.F. (1959). Almost reducible systems. *Sibirsk.Mat.Z.* **7**, 751-784

Coppel, W.A. (1967). Dichotomies and reducibility I. *J.Differential Eqns.* **3**, 500-521

Coppel, W.A. (1968). Dichotomies and reducibility II. *J.Differential Eqns.* **4**, 386-398

Daleckii, Jn.L. and Krein, S.G. (1950). Some properties of operators depending on a parameter. *Dopovidi Akad.Nauk.Ukrain.SSR* **6**, 433-436

Erugin, N.P. (1938). On an exponential substitution for systems of linear differential equations (the problem of Poincare). *Mat.Sbornik.* **45**, 509-517

Erugin, N.P. (1941). A remark on Shifner's article. *Izv.Acad.Nauk SSSR Ser.Mat.* **5**,79-86

Erugin, N.P. (1946). Reducible systems. *Trud.Mat.Inst.Steklov.*

Freshchenko, S.F. and Shkil, N.I. (1960). Asymptotic solution of a system of linear differential equations with a small parameter. *Ukr.Mat.Zhurnal.* **12**, 429-440

Gel'man, A.E. (1957). On the reducibility of one class of system of differential equations with quasi-periodic coefficients. *DAN SSSR* **116**, 535-537

Gel'man, A.E. (1959). On the question of periodic solutions of differential equations of synchronous motion. *IZV LETI*, in Ul'yanova (Lenina) **39**, 285-291

Gel'man, A.E. (1965). On the reducibility of a system with quasi-periodic matrix. *Diff.Uravneniye* **1**, 283-294

Golomb, M. (1961). On the reducibility of certain linear differential systems. *J.Reine.Angw.Math.* **205**, 171-185

Koval, P.I. (1955). On the stability of solutions of systems of difference equations. *DAN SSSR* **103**, 549-551

Koval, P.I. (1957). Reducible systems of difference equations and their solutions. *USP Mat.Nauk.* **12**, 6-78

Langenhop, C.E. (1960). On bounded matrices and kinematic similarity. *Trans.Am.Math.Soc.* **97**, 317-326

Lazer, A.C. (1971). Characteristic exponents and diagonally dominant linear differential systems. *J.Math.Anal.Appl.* **35**, 215-229

Lillo, J.C. (1961). Approximate similarity and almost periodic matrices. *Proc.Amer.Math.Soc.* **12**, 400-407

Lillo, J.C. (1962). Continuous matrices and approximate similarity *Trans.Amer.Math.Soc.* **104**, 412-419

Markus, L. (1955). Continuous matrices and the stability of differential equations. *Math.Zeits.* **62**, 310-319

Markus, L. and Yambe, H. (1960). Global stability criteria for differential systems. *Osaka.Math.J.* **12**, 305-312

Merkis, V.M. (1968). The reducibility of a certain system of differential equations with almost periodic coefficients. *Litovsk. Mat.Sb.* **8**, 101-107

Merkis, V.M. (1969). The reducibility of a certain two-dimensional system of differential equations. *Litovsk.Mat.Sb.* **9**, 755-781

Millionscikov, V.M. (1967). The connection between the stability of characteristic exponents and almost reducibility of systems with almost periodic coefficients. *Differencial'nye Uravnenija* **3**, 2127-2134

Millionscikov, V.M. (1968). A criterion for the stability of the probable spectrum of linear systems with recurrent coefficients and a criterion for the almost reducibility of systems with almost periodic coefficients. *Dokl.Akad.Nauk.SSSR* **179**, 538-541

Millionscikov, V.M. (1972). On the relation between stability of characteristic exponents and almost reducibility of systems with almost periodic coefficients. *Differential Eqns.* **3**, 1106-1109

Morozov, V.V. (1959). On a problem of N.P. Erugin. *Izv.Uch.Zaved. Matem* **11**,283-292

Palmer, K.J. (1979). A diagonal dominance criterion for exponential dichotomy. *Bull.Austral.Math.Soc.* **21**, 26

Romanov, V.I. (1969). The reducibility of certain systems with periodic and quasi-periodic matrices. *Izv.Vyss.Ucebn.Zaved. Matematika* **86**, 70-73

Romanov, V.I. and Harashal, V.H. (1966). Reducibility of linear systems of differential equations with quasi-periodic coefficients. *Differencial'nye Uravenija* **2**, 1423-1427

Samiolenko, A.M. (1966). Reducibility of a system of ordinary differential equations in a neighbourhood of a smooth integral manifold. *Ukrain.Mat.Z.* **18**, 41-64

Samiolenko, A.M. (1968). The reducibility of systems of linear differential equations with quasi-periodic coefficients. *Ukrain.Mat.Z.* **20**, 279-281

Vulpe, I.M. (1972). A certain theorem of Erugin. *Diff.Uravneniye* **8**, 2156-2159

Wu, M.Y. (1971). Some new results in linear time varying systems. *IEEE Trans.Automatic Control.* **20**, 159-160

Yakubovich, V.A. (1949). Certain conditions for the reducibility of a system of differential equations. *DAN SSSR* **107**, 577-591

(iii) References on Stability

Abel, J. (1971). Uniform almost orthogonality and the instabilities of an almost periodic parametric oscillator. *J.Math.Anal. Appl.* **36**, 110-122

Alekseev, V.M. (1960). On the asymptotic behaviour of solutions of slightly nonlinear systems of ordinary differential equations. *Dokl.Akad.Nauk.SSSR* **134**, 247-250

Anderson, B.D.O. and Moore, J.B. (1969). New results in linear system stability. *SIAM J.Control* **7**, 398-414

Ascoli, G. (1950). Osservazioni sopra alcune questioni di stabilita. *Atti Accad.Naz.Lincei Rend.Cl.Sci.Fiz.Mat.Nat.* **8**, 129-134

Bekbaev, S. and Jataev, M. (1967). The investigation of certain critical cases of the instability of almost periodic motions. *Izv.Akal.Nauk.Kazh SSSR Ser.Fsy.Math.Nauk.* **3**, 8-11

Bellman, R. (1948). On an application of a Banach-Steinhaus theorem to the study of the boundedness of solutions of nonlinear differential and difference equations. *Ann.Math.* **49**, 515-522

Bohr, H. (1933). Stabilitet og Naestenperiodicitet. *Mat.Tidsskrift* 21-25

Burd, V.S. and Baberov. T. (1967). The stability of branching periodic solutions of certain systems of differential equations *Dokl.Akad.Nauk.SSSR* **176**, 991-993

Burd, V.S. and Kolesov, J. (1970). On the dichotomy of solutions of functions of functional-differential equations with almost periodic coefficients. *Soviet Math.* **11**, 1650-1653

Burton, T.A. (1966). Some Liapunov theorems. *SIAM J.Control* **4**, 460-465

Bylov, B.F. (1959). On stability from above of the greatest characteristic index of a system of linear differential equations with almost periodic coefficients. *Mat.Sb.(N.S.)* **48**, 117-128

Caligo, D. (1940). Un criterio sufficiente di stabilitita per le soluzioni dei systemi di equazioni integrali lineari, *Atti 2° Congresso.Un.Mat.Ital.* 177-185

Cesari, L. (1963). Asymptotic behaviour and stability problems in ordinary differential equations. Springer Verlag, Berlin

Conley, C.C. and Miller, R.K. (1965). Asymptotic stability without uniform stability: Almost periodic coefficients. *J.Diff.Eqns.* **1**, 333-336

Conti, R. (1955). Sulla stabilita dei systemi di equazioni differenziali lineari. *Riv.Mat.Univ.Parma* **6**, 3-35

Conti, R. (1976). Linear differential equations. Academic Press, London

Coppel, W.A. (1963). On the stability of ordinary differential equations. *J.London Math.Soc.* **38**, 255-260

Coppel, W.A. (1972). Linear systems. *Notes on Pure Math.Aus.Nat. Univ.* **6**

Cremer, H. and Effertz, F.H. (1959). Uber die algebraischen kriterien fur die stabilitat von regelungssystemen. *Math.Ann.* **137**, 328-350

Daleckii, Ju.L. and Krein, M.G. (1974). Stability of solutions of differential equations in Banach space. *Amer.Math.Soc.Transl.* Provendence, Rhode Island

Dauer, J.P. (1973). Perturbations of linear control systems. *SIAM J.Control* **9**, 393-400

Diliberto, S.P. (1950). On systems of ordinary differential equations. *Contributions to the theory of Nonlinear Oscillations. Ann.Math.Studies* **20**, 1-38

Fink, A.M. and Frederickson, P.O. (1971). Ultimate boundedness does not imply almost periodicity. *J.Diff.Eqns.* **9**, 280-284

Fink, A.M. and Seifert, G. (1969). Liapunov functions and almost periodic solutions for almost periodic systems. *J.Diff.Eqns.* **5**, 307-313

Fomin, V.N. (1968). The dynamic instability of linear systems with almost periodic coefficients. *Dokl.Akad.Nauk.SSSR* **178**, 43-46

Gabeleya, A.G., Ivanenko, V.I. and Odarin, O.N. (1975). Stabilizability of linear autonomous control systems. *Kibernetika* **3**, 69-72

Gurtonvik, A.S. and Neimark, J.J. (1969). On the question of the stability of quasi periodic motions. *Diffencialnje Uravenija* **5**, 824-832

Halanay, A. (1962). Asymptotic stability and small perturbations of periodic systems of differential equations with retarded arguments. *Uspehi Mat.Nauk.* **17**, 231-233

Halanay, A. (1966). Differential equations, stability, oscillation, time lags. Academic Press, New York

Hale, J.K. (1965). Sufficient conditions for stability and instability of autonomous functional-differential equations. *J.Diff. Eqns.* **1**, 452-482

Hale, J.K. and Stokes, A.P. (1961). Conditions for the stability of nonautonomous differential equations. *J.Math.Anal.Appl.* **3**, 50-69

Hautus, M.L. (1970). Stabilization, controllability and observability of linear autonomous systems. *Indagationes Math.* **31**, 443-448

Ikeda, M., Maeda, H. and Kodama, S. (1972). Stabilization of linear systems. *SIAM J.Control* **10**, 716-729

Jakubovic, V.A. (1964). The method of matrix inequalities in the theory of stability of nonlinear control systems. *Automat. i Telemeh* **25**, 1017-1029

Kato, J. (1970). Uniform asymptotic stability and total stability. *Tohoku Math.J.* **22**, 254-269

Kato, J. and Strauss, A. (1967). On the global existence of solutions and Liapunov functions. *Ann.Mat.Pura Appl.* **77**, 303-316

Kato, J. and Yoshizawa, T. (1970). A relationship between uniform asymptotic stability and total stability. *Funkcial Ekvae.* **12**, 233-238

Krein, M.G. (1964). Lectures on stability theory of solutions of differential equations in Banach space. Kiev ,186pp.

LaSalle, J.P. (1957). A study of synchronous asymptotic stability *Ann.Math.* **65**, 571-581

LaSalle, J.P. (1968). Stability theory of ordinary differential equations. *J.Diff.Eqns.* **4**, 57-65

LaSalle, J.P. and Lefschetz, S. (1961). Stability by Liapunov's Direct Method with Applications. Academic Press, New York

Levinson, N. (1949). On stability of nonlinear systems of differential equations, *Colloq.Math.* **2**, 40-45

Lillo, J.C. (1960). Continuous matrices and the stability theory of differential systems. *Math.Zeitschr.* **73**, 45-58

Lukes, D.L. (1968). Stabilizability and optimal control. *Funkcialaj Ekvacioj* **11**, 39-40

Lyascenko, N.Ya. (1954). On the asymptotic stability of the solutions of systems of differential equations. *Dokl.Akad.Nauk. SSSR* **96**, 237-239

Markus, L. (1955). Continuous matrices and the stability of differential systems. *Math.Zeitschr.* **62**, 310-319

Markoff, A. (1933). Stabilität im Liapounoffschen Sinne und Fastperiodizität. *Math.Zeitschr.* **36**, 708-738

Massera, J.L. (1956). Contributions to stability theory. *Ann.Math.* **64**, 182-206

Massera, J.L. (1966). Linear Differential Equations and Function Spaces. Academic Press, New York

Millar, R.K. (1965). On asymptotic stability of almost periodic systems. *J.Diff.Eqns.* **1**, 234-239

Nakajima, F. (1974). Separation condition and stability properties in almost periodic systems. *Tokoku.Math.J.* **26** ,305-314

Nasyrov, R.M. (1964). On stability of almost periodic motions in certain critical cases. *Trudy Univ.Druzby Narodov im P.Lumumby.* **5**, 30-44

Onuchie, N. (1971). Invariance properties in the theory of ordinary differential equations with applications to stability problems. *SIAM J.Control* **9**, 97-104

Opial, Z. (1961). Sur une equation differentielle presque-periodique sans solution presque-periodique. *Bull.Acad.Polon.Sci. Ser.Sci.Math.Astron.Phys.* **9**, 673-676

Perron, O. (1930). Die stabilitätsfrage bei Differentialgleichungen *Math.Zeitschr.* **32**, 703-728

Persidskii, K.P. (1946). On the theory of stability of solutions of differential equations. *Uspehi Mat.Nauk.* **1**, 250-255

Putnam, C.R. (1954). Stability and almost periodicity in dynamical systems. *Math.Soc.* **5**, 352-356

Putnam, C.R. (1960). Unilateral stability and almost periodicity. *J.Math.Mech.* **9**, 915-917

Sacker, R.J. and Sell, G.R. Existence of dichotomies and invariant splittings for linear differential equations. *J.Diff.Eqns.* I, **15**, (1974), 429-458; II, **22**, (1976), 478-496; III, **22**, (1976), 497-522

Seifert, G. (1963). Stability conditions for separation and almost periodicity of solutions of differential equations. *Contri. Differ.Eqns.* **1**, 483-487

Seifert, G. (1963). Uniform stability of almost periodic solutions of almost periodic systems of differential equations. *Contri. Differ.Eqns.* II, 269-276

Seifert, G. (1965). Stability conditions for the existence of almost periodic solutions of almost periodic systems. *J.Math. Anal.Appl.* **10**, 409-418

Seifert, G. (1967). Stability and uniform stability in almost periodic systems. *Proc.USA-Japan Seminar. Differ.and Functional Eqns.*, 577-580

Seifert, G. (1967). On total stability and asymptotic stability. *Tohoku.Math.J.* **19**, 71-74

Seifert, G. (1968). Almost periodic solutions and asymptotic stability. *J.Math.Anal.Appl.* **21**, 136-149

Sell, G.R. (1966). Periodic solutions and asymptotic stability. *J.Diff.Eqns.* **2**, 143-157

Strauss, A. and Yorke, J.A. (1969). Perturbing uniform asymptotically stable nonlinear systems. *J.Diff.Eqns.* **6**, 452-483

Stokalo, I.Z. (1945). Criteria for stability and instability of the solutions of linear differential equations with quasi-periodic coefficients. *Akad.Nauk.SSSR Infomaclinii Bguleten* **1**, (10-11), 38-39

Stokalo, I.Z. (1946). A stability and instability criteria for solutions of linear differential equations with quasi-periodical coefficients. *Rec.Math.N.S.* **19**, (61), 263-286

Stokalo, I.Z. (1946). On the theory of linear differential equations with quasi-periodic coefficients. *Sb.Trudov Inst.Mat. Akad.Nauk.USSR* **8**

Stokalo, I.Z. (1961). Linear Differential Equations with Variable Coefficients. Gordon-Breach, New York

Swick, K.E. (1969). On the boundedness and the stability of solutions of some nonautonomous differential equations of the third order. *J.London Math.Soc.* **44**, 347-359

Taam, C.T. (1966). Stability periodicity and almost periodicity of solutions of nonlinear differential equations in Banach spaces. *J.Math.Mech.* **15**, 849-876

Tazimuratov, I. (1966). On the existence of almost periodic solutions to systems close to those of Liapunov. *Izv.Akad.Nauk. Kazah.SSSR Ser Fiz.Mat.Nauk.* **1**, 102-106

Thurston, L.H. and Wong, J.S.W. (1973). On global asymptotic stability of certain second order differential equations with integrable forcing terms. *SIAM J.Appl.Math.* **24**, 50-61

Vrkov, I. (1956). Integral stability. *Czech.Math.J.* **9**, 71-128

Yoshizawa, T. (1964). Ultimate boundedness of solutions and periodic solution of functional differential equations. *Coll.Int. Vib.Forces in Nonlinear Syst.Marseille*, 167-179

Yoshizawa, T. (1964). Extreme stability and almost periodic solutions of functional differential equations. *Arch.Rational Mech. Anal.* **17**, 148-170

Yoshizawa, T. (1964). Asymptotic stability of solutions of an almost periodic system of functional differential equations. *Rend.Circ.Math.Palermo.* (2), **13**, 209-221

Yoshizawa, T. (1967). Existence of a globally uniform asymptotically stable periodic and almost periodic solution. *Tohoku Math.J.* **19**, 423-428

Yoshizawa, T. (1967). Stability and existence of periodic and almost periodic solutions. *Proc.USA-Japan Seminar Differ.and Functional Equations*, 411-428

Yoshizawa, T. (1968). Stability and existence of a periodic solution. *J.Differ.Eqns.* **4**, 121-129

Yoshizawa, T. (1969) Some remarks on the existence and the stability of almost periodic solutions. *SIAM Advances in Differential and Integral Equations*, 166-174

Yoshizawa, T. (1971). Stability for almost periodic systems. *USA-Japan Seminar on Ordinary Differential Eqns.and Functional Eqns.*, 29-39

Zikov, V.V. (1976). Some questions of admissability and dichotomy, the method of averaging. *Izv.Akad.Nauk.SSSR Ser.Mat.* **40**, 1380-1408

Zmuro, V. (1971). Reduction of a problem on stability of solutions for a differential equation with almost periodic coefficients to multi-dimensional determinants. *Dopovidi Akad.Nauk.Ukrain SSR Ser.A.*, 256-259

SUBJECT INDEX